Quotes from the Journals of a Time Traveler

Quotes from the Journals of a Time Traveler

Mikael Kvist

Quotes from the Journals of a Time Traveler
Author: Mikael Kvist © 2026
Editor and translator: Betina Brandes
Cover: AI assisted
Published by: SPEKTRA-ART Förlag
Web site: spektra-art.se
Edition: 2nd edition 2025
Printing: Libri Plureos GmbH, Hamburg, Germany
ISBN: 978-91-88422-48-4

Contents

Introduction

I have had my documents stored in several different hiding places. In fact, they have been in the same place, but in different times. It was a necessary action I had to take in case I was killed and lost the ongoing time war. I was killed. But I survived, albeit in a state of total amnesia about who I was, what I had done, my abilities, etc.

Since there were no things from my time travel, I could not remember them. They had not happened yet either. In severe amnesia, you don't even know that you can remember. That's where I was. In my home, I walked into a mysterious library that was filled with books and binders. I saw that they had my handwriting but I couldn't relate to the contents because I couldn't understand what was written there.

Thankfully, I did not throw it away. I kept the door closed instead because the room made me dizzy and mentally ill. It was only after a year of flashes of light and headaches ushering in every moment of recollection that I dared to open the door to the room again. Then I could read what was written there. I could understand, and at the same time not. It told me incomprehensible things I had been able to do before. I began to remember episodes, but the abilities were still locked inside me, inaccessible.

Where should I start to remember the abilities? It felt unreal, totally unreal. But my recurring memories said otherwise. I had done everything I had written about. There was only one thing to do. Sit down and try to piece together how I was once a portal walker.

Nothing is as slow as getting memories back. In the case of total loss, you can't even know that memories are missing. The memories you have are compressed without holes. Suddenly out of nowhere episodes and even abilities come back. This is what happened to me recently.

The desk, which I know is empty for a reason, had documents in it that I had of course totally forgotten. They were my travel notes. How the notes could suddenly be there has to do with me not remembering the drawer as empty.

If I remember it as empty, I instead act as the observer who confirms and verifies the emptiness of the drawer every time I open it. My Schrödinger's cat is about whether or not there is another document when I open it.
So my trick to get the notes back is not to remember the drawer as empty. A fantastic trick to be able to fish up memories. I don't know how many I can retrieve. I take the notes as they come back.

I have an advantage and that is that the more notes I bring home in my time here, the more awakens within me. Memory is what connects times and old memories of other times. With a new glimpse inside, I open the drawer carefully, almost as if squinting. I must not think, I must take it as if it is the most natural thing in the world that this is where I put them.

The times I succeed in retrieving notes, I have tricked myself into being empty-minded. The times I have failed, I have chosen to forget so that they do not color me.

How can the observer determine if notes exist or not? Well, that's in the journals you're getting quotes from here.

Quotes from my journals

Me: It's an honor to meet you.
I am a time traveler.
Do you have a minute to spare?

He asks: Have we met before?
I answer: Yes, several times,
in the future.

Reverse cause and effect exists.
We had to meet
because of the effects.

I haven't said these quotes yet,
I have brought them here
from the future.

If empty space, vacuum, were
really empty,
we could travel infinitely fast
from one place to another.

Light would have no speed
and time would end.

Where I enter in time,
I enter into consciousness.
Therefore I enter into your mind
and not with a flash.

You will be thinking of
something else
and will not notice me appearing.

I, through time travel, always
enter unnoticed
and exit unnoticed.

You cannot experience in time
when I appear,
only when I appear in your
consciousness.

When I appear in your
consciousness
I change time
and you believe that I was
there all the time.

It's time that makes space.
I don't travel through space to
reach you,
it's too far.
I travel through time in an instant.

Space is not the distance,
time is.
Space is the result of your time.

When times collapse, realities collapse.
There are many times open for,
but few make the passage.

These were diagnosed with psychosis.

The thing that protects you from collapse
is a mental blind spot.
It makes the mind contract time
before and after the opening.

The blind spot gives you a
smoothing
micro blackout.

The mathematics of existence

He: You know that my colleagues and I have been working on finding the equation that describes the very beginning of our universe. And if we do, we hope to find the solution to the theory that explains all of physics, not in parts like general relativity and quantum theory do.

Me: Have you tried starting from the beginning instead of ending there?

He looked puzzled and that suited me just fine.

The Mathematics of Existence
not yet discovered
starts with a 1
and a 0

The fundamental existence
is binary.

If you exist
there is also a zero
If you don't exist
there is also a one.

The existing
and the non-existence
confirms their coexistence
as 2 that are 1.

When One looks into the depths of Zero
it sees all that One is not.

When Zero looks into the depths of One
it sees everything that Zero is not.

Eternal

When both look into the depths of the other
they fall
eternally unifying

in what can be called love.

They are inseparable.

They are 2
that are One.

I say again,
The Mathematics of Existence
not yet discovered
starts with a 1
and a 0.

You can trust
that the rest of creation
is mathematical,

in detail
and as a whole.

He: Where do we go from here? I mean I could probably start with your suggestion here, albeit a little too poetic for my taste but I understand.

Me: You yourself have said that math is calculation and logic, and I can tell you that it's a wonderful tool in your own world of strange time. All you need to add to your equations is for the observer to make the equation true.

He laughs and says: You continue to truly amaze me! I must say that I am very doubtful about you, very often, but suddenly you make me unexpectedly enthusiastic. You haven't told me how, and it still makes me hesitant but...

Me: I'm not here to solve, that would be to interfere, sorry, I'm just here to inspire. And I can tell you, based on your reactions, that you already have an understanding. Not yet formulated, by that I mean mathematically, but it will fascinate you so that you can never stop exploring.

I smiled for a short while before continuing.

Me: You don't yet know that math is not just logic and calculating but includes an observer. Yes, you can deduce with logic that if A equals B and B equals C then logic says that C equals A. But don't forget that it is the observer that makes the equation work. And now for the leap. If you can also fit the observer into math, you will find the God Equation.

I did not continue. He was definitely not ready for more yet. I had taken a step too far. My credibility fell to a new low. This will be a long road to understanding. What I just told him was my first introduction to Aristotle's Prime Mover.

Perhaps I should have tantalized him instead by continuing on Euler's Equation? Too late in any case. And maybe I should have replied to him that poetry can actually be the inspiration for mathematical beauty, quoting what math professor Keith Devlin said: “Like a Shakespearean sonnet that captures the very essence of love, or a painting that highlights the beauty of the human form that is much more than skin deep, Euler's Equation reaches down into the very depths of existence”. And added that: So does Prime Mover.

Does the universe have a purpose?

He: I know this may sound theological but with your perspectives I wonder if you can shed some light on this. I mean it's a common question, but not so often acknowledged. I swear it's been on the minds of every physicist, whether they admit it or not.

Me: I think you know the answer. Everyone asks that question because they instinctively know the answer. Time creates distance between events. This distance gives consciousness the possibility of experience.

He: So you mean that time has a purpose?

Me: Yes, it gives consciousness the possibility of experience.

Time makes consciousness
experience
itself
and also become itself.

This makes time a journey.

When the eternal soul,
a unit of consciousness, thus You,
travels through time
it gains the experiences.

It always returns to itself.
This creates loops of time in time.

What makes the Hero's Journey,
a theory by Joseph Campbell, universal?
It is the journey you find
in every archetypal story of a hero.

The hero makes a journey
out of his limited experience
and comes back, transformed.

This is really the story
of how the Universe
gets its experiences.

Every soul is born to be a hero.
You knew this.
You can feel it.

The question is,
will you follow your mission?

A Universe that does not experience itself
does not exist.

An observer who experiences in the
Universe
makes the Universe exist.

Everyone is a co-creator.

Gods did not understand that everyone
is a co-creator.
They knew what they saw.
They saw competition
for their self-proclaimed position
as omnipotent.

Gods in this perspective are beings
who became creators
before they became self-aware.

They lacked that experience.

The hero always travels
towards the unknown and
unexplained.

By the known, the unknown is
perceived as a threat.
A consuming hole in all that is known.

The hero who seeks the unknown
is seen by the All-Knowing as a
personification of the threat.

He who disturbs the order
is seen as evil
against the order of the Almighty
and must be stopped,
with equally monstrous force.

The hero will inevitably
face the monster.

For the All-Knowing
the hero is the accusing evil.
For the hero
the All-knowing is the challenge.

Is this not contradictory?
The All-Knowing should know,
but what the All-Knowing does not
know,
it does not know.
It only knows the known.

That the Hero unknowingly knows cannot
fail.
The Hero is the will of the Universe to
travel
to experience and also to return to itself.

Time made the journey possible.

The hero made a time loop
by returning to himself.

The time loop confirmed
the existence of the Universe.

When time travel is not possible

He: I was just reminded that Stephen Hawking held a time travelers party, in 2009 I think it was. Hawking held an invitation-only party and provided balloons, champagne and snacks for his guests, but he didn't send out the invitations until after the party was over. I think you know about that...

Me: You go ahead.

He: In the invitation, Hawking gave the date, time and GPS location of when and where the party had taken place. And as you know, no one showed up. Based on this, Hawking could claim that "I have experimental evidence that time travel is not possible". And now to my question. Why did no one show up? I'm pretty sure you're a time traveler, but you're only convinced the day I travel in time myself. So what do you have to say about that?

Me: I'll start at the uncomfortable end... Hawking is much revered in your time. I agree that he did remarkable scientific work but he had his flaws, as almost all physicists do.
I looked at him. His face was easy to read. I continued.

Me: It proved nothing. It can't even be called an experiment. But just a way to prove his own point.

And if someone were to show up, do you think they would make themselves known? My answer is no, and I'll explain.

I paused because I had his attention and I wanted him to listen carefully.

Me: If that event, on June 28, 2009, was indeed an experiment intended to prove something, I ask; where was the preparation?

He: He sent out the invitations after the event, but he couldn't do it before the event.

Me: Wrong! He predicted the Hawking radiation from black holes, so he could have predicted the effects of time travel. We use micro wormholes when we want to compare compatible spacetime curvatures for black holes.

He almost lost his breath but managed to say: Hey, you really say things that make one's jaw drop sometimes. Sorry, please continue.

Me: First of all, all wormholes have similarities with tunneling. It would cause the surrounding air to suddenly become superconducting, in other words, my trip would have locally and temporarily affected the readings on an ion meter. No such meter was used. Nor any measurement of any radiation in the full spectrum. If he would have at least used a full spectrum video camera, you know the kind that has its infrared and UV filters removed to limit the camera to recording

only visible light.
I looked at him and he was paying attention. I continued.

Me: At any covert military installation, we would encounter many devices, such as ultra-high UV light detectors and anomalies in laser beams. Well, you should know that we do not approach them. They are not of interest to us as the military seems to think... So do you understand now why neither I nor anyone else phased-in to his time and matter frequency that would have made us visible?

He: Why me?

Me: You are open minded.

He: I thought Hawking was open-minded, but I realize he wasn't. But why is that so important?

Me: A quantum-influenced observer bias.

He: You mean there's an observer effect

Me: You, the observer of this universe, are the master in that you are the master-confirmer of the manifested state of all quantum particles. You create and maintain your universe and I just slipped in through a gap in your reality. That gap was in your mind. One uncontrolled moment after another of absolute transparency. A blank page has a blank surface, your mind was a mind of open space that I could enter through.

He: Was it easy? I wonder...

Me: You don't know how many times I've tried. And when you find out by recognizing these moments, I will have left for the last time. From your time frame, I add.

I saw that I touched him deeply. He would start to remember. The feeling first, someone close, familiar, dear, and then the moments when I was so close to him.

Scientific arrogance

It was with his quizzical look and his very special smile that he once again couldn't help but come back to Stephen Hawking's experiment of inviting time travelers to his party. I could tell that he wanted to take the subject a step further. He wasn't finished with certain questions and I think he was searching for them, because he was repeating what both he and I had said earlier.

He: I understand that he wanted to prove that time travelers do not exist and that puzzles me. I haven't seen him like that... but I realize you must be right.

Me: You know what I call it? I paused to give my words the weight of being well thought out. Scientific arrogance!

He coughed and with a smile he managed to get out: Those were strong words I must say. Not many people dare to question Hawking. Nor do I, for I have seen him as one of the giants on whose shoulders I stand.

Me: There is a difference between what he showed and what he actually demonstrated. And it's about scientific rigor. Before you carry out an experiment, you have to be extremely careful about how you plan it. It's easy to introduce both quantitative and qualitative errors, in addition to not meeting validity. The quantitative figures were zero visitors but... This was a psychological experiment as it involves humans and human behavior. I therefore argue that he did not have the capacity to draw the qualitative results. I argue that this should have been done as an interdisciplinary work including both the faculty of psychology and social psychology, as it involves both internal decisions and social interaction.

I paused briefly to allow his thoughts to settle before continuing. I wanted him to be fully on board with my reasoning.

Me: Validity was not met because the question construction was not well thought out. And reliability requires instrumental experimental design to be repeatable. So he would have needed to hire a technician skilled in detection.

He: Excuse me for interrupting, but what instruments could be used. I know you mentioned something about that last time, but now you have made me more curious. Quartz clocks set up in the room for example?

Me: Yes, but that only works for longer openings. I enter through a gap. Besides, clocks self-normalize as soon as the opportunity arises, the nature of time is such. No, you could easily measure deviations in magnetic fields, a short pulsation within a few Hertz or so.

In case of high phasing-in, you could use a

capacitive proximity sensor or pass control with analog recording laser deviation. In other cases, I suggest measuring relative ionizability as ionization occurs at the elementary particle level in the form of electrons and this is influenced by local wormholes to, without achieving the necessary excited state, tunnel to leave the object. I can describe on another occasion what I would put your engineers to work building.

He would laugh in his jovial way, always smiling. I think it is his smile that makes me enjoy telling him. He smiles because he enjoys our conversations.

He: Sorry, I interrupted you, go on. Where were we?

Me: We were at the lack of validity and reliability. I continue. The shortcomings in Hawking's case mean that it can only be concluded that no time travelers were detected by the methods used on that particular occasion. So the formulation could at most say: In my experiment, and with the conditions on the occasion, no visits by time travelers were detected. No instruments were used to record typical signatures of entry into time. So there were only eyes to record with and these were not even trained. You need insight into how humans avoid the surface where the entry is on its way, because they perceive it already occupied. In their mind, there is already someone standing there, occupying the surface of the floor.

His face is so transparent. He gasped and I realized that he must have remembered a situation where he had walked around nothing on an open floor. His eyes said there was a lot going on in there but I chose not to read him.

He: You seem to have read up on this. You have a degree in psychology it seems?

Me: The biggest challenge in time travel is anticipating interaction with other humans and drawing conclusions about where my action will direct them in the moment and in the future I come from. Study and knowledge of interaction is a necessity.

He: What do you think Hawking's mistakes in this experiment have to do with? I admit that you've knocked the foundation I thought I was standing firm on, but at the same time I know that he is an unquestionable authority on astrophysics.

Me: Hawking's scientific arrogance has to do with his relationship with God.

He: But I didn't think he had any relationship with God!

Me: Well, that's when you do, very much so.

No time for God to exist

He: Hawking said that he could prove God's non-existence by showing that there was no time for God to create the universe at the moment of the Big Bang. He says that time started with the Big Bang and therefore there was no time for God to create the universe. That is, there is no God. I think there is a flaw in that argument, because Einstein's theory of gravity breaks down at the moment of the Big Bang. If you start from string theory then...

Me: Stop there. You don't need to move on to yet another promising but untested theory of the universe to see the errors...

He lost his breath and in the meantime I made sure to finish before he got on to his favorite topic.

Me: I can give you hints on how to proceed with string theory, but later. Let's analyze Hawking a little more thoroughly first.

I managed to hold his horses that just wanted to go for a quick thought ride across the green meadows of string theory.

Me: Let me put it this way. We can agree that you can't make steam engines fly.

He nodded, but didn't smile yet. I continued.

Me: We know that steam engines can never fly because they are too inefficient and they are too heavy. Especially with the technology that was available at the height of steam engines. So what I'm saying is that your science is limited by your technology and your accumulated knowledge.

I looked at him with a straight face.

Me: You are right now on a steam train of your time that will never fly. The tracks, and thus the road, are paced by current assumptions. The only correction is when someone stops at a station and manages to switch tracks to a slightly different direction. You are still bound by your technology and your limited vision.

Now understand this. The Big Bang is an assumption that is not scientifically proven. A sudden explosive creation of matter from nothing is yet another assumption based on the first. You don't have the understanding or the technology to understand and build a working understanding, based on what you know. From my point of view, this is not science, but advanced guesses. What makes it advanced guesses? We use math to prove that assumptions have no validity. Math can always be used, even to predict. But I predict that you don't understand how to build a flying steam engine, nor how to build a real flying machine with current primitive knowledge.

He: Are you saying that we lack both the technology and the knowledge to understand or even figure out how the universe is created?

Me: Yes! Just as you once predicted that trains cannot go faster than 35km/h because the air would be sucked out of the train and humans would die. To make a brand new discovery, stop at the nearest station and look closely at how narrowly tracked your path is.

I paused for a short while while he tried to make some sense of it.

Me: You think you understand what time is, what space is and mostly what energy is. But you definitely do not! Imagine what we know. We know how matter is shaped by thought, how time is constructed by thought. And with thoughts about how the universe works, we don't have the energy to do the job. We use our thought.

He: That's amazing. If you're right, I don't know what to say.

Me: Don't say anything. Don't fill the infinite space I just gave you with all kinds of thoughts.

A creation is not a moment.
A creation is time as a whole in an instant,
all the time.

You can see it as the entire timeline of
an object is inserted in time.
A created object suddenly exists with a future
and a past.
A second ago it was born
with an age of billions of years and billions to
come.

This is why you never see a creator.
A true creator is beyond time.

Science is the understanding of cause and
effect,
in other words, events in time.
Creation is an event beyond time.

He: Now, if I were to dare to take a completely different approach, purely as a thought experiment, where would I start? Take our planetary system as an example, I would argue that we have the tools to explain without involving any God. After all, we have our laws of nature to lean on.

Me: Okay, you asked to lose the ground beneath you. The unfathomable abyss of time is far more terrifying than that of space, for it is time that builds the depth of the abyss.

Creation never violates any natural law. What is created is part of the existing universe. The universe has a will. I think you understand after this analogy. Imagine that your moon did not always exist. All life would be different, including the weather. There would be no clear nights.

Imagine that 40 000 years ago a moon appeared in your sky. The humans who lived then would see it come into being and tell legends about the event. Imagine that the moon is not yet created. We have not yet arrived at that future, but it is approaching. Imagine that I will be in that future when the moon is created. I would not be able to visit the moon before it is created, because it would weaken the original to be able to come into being by its own existence because of my feet on its surface. But once the moon is created and inserted into time 40,000 years ago, it will have a history of over 4 billion years and as many to come.

And without the moon, life would not have arisen. Nor would I, who helped create the moon. Its perfect diameter, the same as your sun as seen from Earth, its perfect orbit around the Earth and its unnatural light gravitational influence at that distance. The lightness and the earthquake sound that suggests the moon is hollow.

When you see the moon, it is a created miracle that is too perfect to be true, just as all the constants of nature are perfectly balanced to each other to cause stars to form, life to emerge and persist for billions of years.
When you imagine the grandfather paradox, it's nothing compared to the real paradoxes that obey the laws of time.

He: Well, what shall I say after this thought experiment? Because that's what it was, right?

Me: Of course it was a thought experiment, to make you realize that you lack the knowledge to pronounce on God, time or the process of creation. Give yourself time and you will find what I am talking about. And if you wait long enough, I can show you that this was a thought experiment, by taking you to the moon.

I paused and smiled at him.

Me: But that will be after the moon is created!

Reverse causality is purpose

He: You have mentioned many times the characteristics of time, and also so generously shared your own experiences that I wanted you to elaborate more on the subject.

I nodded my head as to say go ahead, I'm ready.

He: You have mentioned reverse causality, for example. You have also mentioned in passing that it is the scientific understanding of reverse causality that would connect science with religion. Could you elaborate on that a little bit... I've increasingly come to realize that you have certain insights that are, well, undeniably anachronistic and to some extent prophetic. Yes, you have made me curious, tell me.

Me: First of all, we need to revise some basic assumptions, the speed of light for example.

He laughed in his soft way. I could tell he was thinking, or rather saying with his laughter, that that was a pretty big revision. I didn't comment, I just smiled in agreement.

Me: And I want to clarify, I didn't say religion but spirituality. Hold on, this is not questioning any available data, nor any relativistic theories, not at all, it is just expanding and broadening the horizons of possible new quantum interpretation and easier understanding. It will make the quantum world more understandable.

He: You are suggesting here that if we revise what we know about light, we can unify relativity and quantum physics. That would be a possible solution to the dilemma called the God Equation. This is no small ambition. Einstein worked on this for thirty years and as far as we know he did not come up with a solution. And I have been working for many years now on the basis of string theory to get there. I think that in this area we have come a long way.

Me: But you, non of you, have gotten where I want to get. Are you along for the ride?

He both nodded and replied eagerly.

He: Of course. These are my favorite areas. If you can give me a hint, I'll be happy to integrate it into my work.

Me: Even if it means starting from scratch?" I said in a humorous tone.

Of course, he replied and I could see in his face that he took it as a joke, which he should. But not that this would be the case. He hadn't touched on his mathematical framework and I didn't want to. I had avoided it so as not to have to say that I can't answer that question because of time laws, it would sound like a lame excuse. I could only hope that our conversations would get to the point where I could explain that I was only allowed to leave behind sensory impressions, not any objects. Mind builds strong time but objects weaken the fabric of time. And equation, complete or in progress, is an object in time from a time traveler's point of view. I turned my focus to him and the topic he had requested, about reverse cause and effect.

Me: The speed of light is the highest speed anything can have. It's the same for everyone and that causes the space to curve instead and our clocks to run differently depending on the speed we travel at. So far we agree. But now you have to hold on again.

I say that light has no speed. And if we start from that, you (I) start and have to build a new physics that has hair-raising consequences with suddenly incorporated quantum physics in our size and scale. Suddenly you can see that there is no magic limit on the size scale where the quantum level collapses into what we call reality. The laws of the quantum level have never collapsed into our reality. Its laws are still there and so is the influence of the observer.

I looked him in the eye to see if he was with me and he was. He was immediately full of paradoxes, unsolvable questions but also thought expansions in different directions. I chose not to condense his thoughts here but to write how we continued my thread in this.

Me: Cause and effect only exist when we have time. Likewise, we have space only when we have time. A photon traveling at the speed of light has no time. As relativity says, if we traveled at the speed of light, no time would pass. But on the other hand, it would require infinite energy to accelerate to the speed of light. As a parenthesis, I can say that this is why none of us accelerate to have the speed of light.

I looked at him and saw his question and answered: Yes, we have the speed of light in that we are timeless and we are timeless in order to enter at the desired spacetime. But we normally phase ourselves in so as not to give rise to a flash of light around us. I continued.
From the photon's point of view, it doesn't take eight minutes from the Sun to the Earth. It hits your eye the moment it is born. And I dare say that a photon that does not meet its confirmation that it exists (the eye of the observer or the matter of the hitting object), that photon has never been born.

He (chuckling but not dismissively): Well, that leaves the question of how to build a sensor that can detect the photon without interacting with it. I don't see how that could be done. So even if you are right, I don't see how it could be proven.

I looked at him with a soft smile because I liked that he liked challenges. And I smiled in advance at my answer to him.

Me: It's already been proven but in reverse.

He: Now you are making me curious. It seems like I've missed something here.

Me: Yes, that's not so strange because it's not your field of research. These days there are almost watertight barriers between scientific branches. Yes, I blame funding and support as the reason for wanting to keep your boundaries. I guess you heard struggle for funds for research. But the future belongs to interdisciplinarity. The answers are in a new branch, quantum biology.

The leaf on the tree receives the photon, converts it into energy for its growth, cell division and more, which is natural. But it is only when you study the path of the photon at the cellular level, that you can realize that the nucleus for the energy conversion in the cell itself is so small that the probability of the photon hitting this tiny nucleus in the already incredibly small cell is too low.

So, why does the photon hit? Because the photon is not born unless it is received and becomes part of the tree's organism and, to some extent, the tree's consciousness. But let's leave that last one aside, let's leave it as a separate question. The statistical calculation says that it is too unlikely that the photon hits the right spot. This study does not mention, but points to, reverse causality.

So let's jump a bit more. I can come back to quantum biology later, but let's go back to the wave nature of the photon. I know it's starting from scratch, but then you'll see how I change the whole quantum physics with a fundamentally revised interpretation of the experiment. In experiments with one light source and one narrow gap (slit), we get a normal distribution curve of the light towards the surface it hits. In the same setup with two slits, bands of light and darkness appear on the receiving surface, depending on whether the waves amplify or cancel each other out. So far so good.

But if we let just one photon through at a time, the photon knows whether there is one slit or two because the distribution curves will be different. Although one photon at a time with two slits can only travel through one of them, it interacts as if there were another photon. From this comes the elegant explanation that the photon with some probability is actually traveling through both slits and we have probability waves. I call this a convenient solution, as convenient as explaining the universe by saying that a new universe arises for every choice we make, The Multiverse Theory. Let's not go into that now but stick to the revised interpretation.

The photon hits at the moment it is born. I say light, meaning photons, has no speed. The speed of light is something that is created by the observer, the consciousness of the system. And as I mentioned, time is necessary for consciousness to have experiences.

Now to the exciting part. If the photon doesn't exist, if a receiver doesn't exist, it implies reverse causality. The photon hits because it has a target, a purpose. The photon does not travel, the photon is born when it has a target.

It was not often that he closed his eyes to his thoughts but now he did. I continued.

Me: This will change the whole of quantum physics, if you start from the single revision that light has no speed! You will fall into a universe that has no one-way cause and effect. All yardsticks will suddenly lack validity and in some cases be reversed. And if you can reconcile this with the fact that everything in the universe is consciousness, you will understand time as a fabric, not a line, and the complex relationships we time travelers have to deal with.

As a time traveler, I can say that if the somewhat facile explanation that a new universe arises every time I travel in time, I wouldn't be bound by all these complex laws that govern time as a fabric. And that fabric is formed by reverse causality. It is a fabric that is without time and with a purpose, the purpose of forming itself. The photon knows this. The photon has the purpose in front of it when it is born. And you become the wandering observer who, through time created by your brain, gathers the experiences that form the universe.

Dark matter is spiritual

Dark matter is not darker than your mind.

He often laughed at what I said. It was laughter that made him feel refreshed.

He: Okay, give me your take on dark matter. I sense you have something clever to say about it.

Me: Dark matter doesn't exist...

He: Somehow I knew you were going to say that.

He smiled and squinted at me, waiting for more. He got more.

Me: If I remember correctly, I can't always keep track of the numbers, and they are different on each visit. Only 5% of the universe is matter that we, uh you, can detect. The rest you have labeled dark matter and dark energy.

He: It hasn't changed that much since you were last here, has it? It's mostly been corrections to more accurate values.

Me: Well, when Hubble came up with his interpretation that the universe was expanding, there were no cosmological problems and his conclusions were obvious for his time. He did the first pioneering work on measuring distances. The problems came later... I think you will see me as older, it will be easier to think that I was here even before you did your first experiments for the Science Fair.

I liked the way he smiled at the thoughts I gave

him. His mind was stimulated. I moved on because his mind gave me an open field.

Dark matter is a mere notion that has become an established concept and is now being tried to prove. But there is not much difference between that and impetus. If you remember, I mean if you have read, Jean Buridan and how since the 14th century impetus was the leading and unquestioned theory for the orbit of both planets and cannon projectiles. After all, a cannonball fell down, straight down, when its impetus was over. It was even considered that the theory did not need to be proven because it was the dominant concept on the market.

This was repeated with the presence of the theoretical ether. The ether was created out of a need to explain. We wanted to have a fixed point in the universe after Newton threw us into explained elliptical orbits in space. The choice was to try to find evidence and explain it with ether rather than look at what it was, until Einstein explained space-time in a new way.

We are facing the same repeated situation. Dark matter has become a concept that people want to find evidence for. But it will work just as badly, because dark matter does not emit, reflect or absorb ordinary light. It is a fancy way of saying that something exists, just as little as the ether with the same properties, even though it does not exist. The most extreme is to reinvent a kind of ether called quintessence, after the 5th element of the Middle Ages. I promise to come back to this, but first I would like to close the issue.

Science has evolved, but humans' minds have not.

Attitudes are genetic.

Everything is different for me.
I have seen how attitudes for centuries
not changed.

A more advanced base says nothing
about the minds that now own it.
They have always identified with
their base.

I have no base of any time to stand on.

I can afford to see the cracks more than
the strength,
and must.

The day the cracks of reality become your
friend,
you won't fall through them
helplessly.

You become a space speleologist.

I will answer you before you ask.
The answer is to always return to the
crack,
not fill it with an accepted concept.

Me: Just giving a name like dark matter to a patch of the crack makes scientists try to prove dark matter. I say again, go back to what gave rise to the misleading concept. Only then can you sense the real answer.

The problem was that galaxies were rotating too fast and would throw out matter and not hold together. What holds galaxies together is something that cannot be explained by the matter they have. Had galaxies at their periphery rotated slower, physics would not have the crack.

I think you're ready to take the big leap now, into the serious unknown that has been known for millennia.

He got excited and there was some necessary chatter before he was ready to finish listening. I smiled too, can't help but like his personality.

Me: How about changing Planck's constant and all that goes with it. The smallest building blocks of matter can only interact with building blocks of the same kind, that's obvious. But no one has tried to see how other values of the constant would work and interact. The interaction could sometimes be partial between different types of matter, but for the most part without being visible. Such matter would not emit, reflect or absorb light.

What about gravitational influence from this invisible matter? It would exist and would vary depending on how many such planes of parallel existence exist in the area.

The knowledge of these parallel planes of existence is thousands of years old. What I have said is that what you call Dark Matter is spiritual.

To understand us, you need to understand our time. Change Planck's constant and you change everything. You get a different speed of light and a different speed of time.

You've already understood that I'm from a different time. Now you understand that I am also from a parallel existence.

We had a break with scattered thoughts that I gathered together, straw by straw. Disordered yet collected, so it gathered his thoughts in a similar way. A disorganized bundle but all in his arms.

Yes, a merger between science
and spirituality
is inevitable.

When your future science
has caught up with
your spirituality from the past,

all planes are open
again.

Many of us have visited you.
The interpretations have been equally
numerous.

Some of us came flying in machines
crossing the boundaries of dimensions
and
the boundaries of planes of existence.

He: Is there any evidence for that? I'm not asking as a skeptic, but more as a curious person.

Me: There is plenty of it! As I said, don't be blinded by concepts like UFOs. They are actually dimensional craft. Because you didn't think we flew through space to get here, did you? That's way too far!

He laughed that satisfying, soft laugh again. I continued because I knew he'd be eager to hear more, without interrupting with questions.

Me: We're traveling through time instead, it's kind of faster that way because it doesn't take any time away from the one I have.

We smiled in agreement and chuckled together at the fun of the words.

Me: By traveling through time, I can choose how I land, which Planck's constant I choose.

I can also choose where in the chosen time I land here, thus as a time traveler. That is the hardest part.

But let me return. Outside this reality, we can accelerate to incomprehensible speeds. We are not dependent on moments of inertia. Entry and exit can occur in two ways. Either we use the material Doppler effect and you see the light of the ionized field around the craft change color to eventually slip out of the visible octave. But we can also disappear to a point. Don't ask me to go into it but our size can "shrink" and we disappear out of a point

He: But if I understand correctly, you did not come in a craft?

Me: A craft is a possible way for anyone to travel in time, planes of existence and space, without having the knowledge in the body. I myself am a portal walker.

You cannot travel in time without also
traveling through parallel planes of
existence.

It makes us spiritually aware.

As a time traveler through parallel
planes of existence
I can say with this,
there are no spiritual planes
that are not also of matter.

It is matter that gives the plane of
existence
its characteristic properties, laws and
possibility to inhabit it.

The definition of a spiritual plane of
existence is
matter of a different kind
that vibrates outside this available
spectrum.
Usually in frequency higher than that
defined by
Planck's package.

Yes, there is more.
Types of matter you can't imagine
and laws of nature based on them.

That galaxies have a flat motion, is
inexplicable.
That the Universe has a flat geometry
is unlikely.

What are the chances of that?
Other than that current cosmology
overlooks the most obvious.

The Universe is shouting the answer
in your face.

Answer the universe
with an open question.

Do not defend
by reinforcing agreed concepts.

The invisible can be seen,
not only as apparitions and
miracles.

Like me, you can travel there.
Or as you may now dare to
count there.

Spirituality has preceded.
Science comes after.

An insight lights up like a sudden light
in minds.
A context lights up overwhelmingly
in hearts.

He (with a puzzled smile): Dark matter is spiritual. I will never forget those words!

I did not answer what I was thinking. Unfortunately, you won't.

A time travelers universe

I had his confidence in a standing invitation. It warmed my feelings towards him that he had confidence in me. I think he now recognized me, though without direct memories of his youth, and made me something of an older relative or rather patron.

To aim and arrive in time is not easy, at least if one is to carry out a series of visits. Time literally has a will and a mind of its own. I am good at following it, but I could land inappropriately. An inappropriate landing is to end up in the wrong chronological order for him. Another is to land in a room he has just been in and knew was empty, or at least thought he knew. Such things can scare the wits out of anyone. I happened to return to a room a friend had looked for me in and found it empty. When I came out of there shortly afterwards, she was so terrified that she not only screamed, but collapsed. Her legs literally could not carry her.

With him I tried to make my presence felt. I had talked about how it can be sensed by the ionization in the air, among other things. The hairs can stand up. And raise his hair I did. I'm telepathic so I heard his question from where he sat at his desk.

Me: Sure, it's me!

He: You're just in time for breakfast.

Me: And I have breakfast with me!

I held up the bags. Our moment was off to a great start! Now he would let loose with questions without inhibitions. And I would be able to take him on time travels I had made in the universe.

He'd have an inspired session first thing in the morning. And breakfast had to come when it was over. We set out the contents of the bags on the kitchen table. He was hungry, in both senses of the word. He piled into the buffet of answers he hadn't seen me unpack. I knew his hand of questions was hungry too, and I suspected it had just what he wanted with his bagel and coffee.

He: How far out have you been? I'm thinking you've visited places in space too. Can you tell me something I don't know about the universe.

Me: I hope you like pancakes too because you're going to have to swallow this with at least one! At least if you just want it served without me giving you the mathematical recipe for the pancake of the universe.

He was in a good mood and we laughed together. I held back a tear, as if some small piece of debris had gotten into my eye. I blinked a few times in an extra-obvious way.

Me: Okay, let me give you a menu. I'll take this

extra plate. A pancake is flat. Who wonders about that? No one. Yet it's one of the greatest cosmological mysteries, why the universe is flat.

I tucked into some bacon, which I had cut while I was talking, along with the egg, which was dripping precariously. He filled me in on the spherical shape and the horse saddle shape and how really any shape would have been more natural for the universe. I nodded, chewing, because I was hungry too, I realized.

Me: So if we were to look at a flat universe, it's almost an impossibility, because any deviation from flat in the beginning would have escalated to sharp deviations in the present. (I paused briefly for a sip of tea from the paper cup.) But if the Big Bang never happened, then the flat pancake is obviously the outcome. Even if the universe had a beginning, it wouldn't necessarily have emerged from a singularity at an infinitesimal point out of nothing. I hope to show you how singularity surrounds us and that 0 is greater than 1.

I didn't expect to hear him talking with a bagel in his mouth, but of course his eagerness drowned out the bagel.

He: Are you saying that the Big Bang never happened? Uhm, as we mean according to the Standard Cosmological Model (Λ-CDM)?

Me: Of course. I've skimmed through too many unnecessary publications that variously compensate for observed facts. I would say that in astrophysics over 50% of what is published is completely wrong! Almost everyone bases their statement on the accepted facts. New eyes are needed on existing facts, because even new facts are adapted to the collectively accepted model.

He didn't cough. Although I would have expected him to. He took it for what it was, an intellectual menu for breakfast. After a few exchanges of ideas, I continued.

Me: To understand why the pancake is flat, you need to understand how something is created from nothing. But I promise we'll do that another time. It assumes that; sure the universe came into being but not the way you think. Everything came into being all over the surface at once and quickly it became "solid" pancake with its little almost invisible bubbles.

And then it came. The Red Shift. It's funny how I know what follows what. It's a necessary observation as a time traveler. But I had discovered that it wasn't always logic that governed the sequence of thoughts in an interaction, especially across time gaps like this one between us. It seemed that time itself had more will to direction at greater distances as well as intense encounter. It was precisely this and that I found the knowledge, though physicists would call it information, in time itself. Time was or bore consciousness.

The insight I was to share with him was about how consciousness builds time and thus space and that the speed of light and the formation of matter is subordinate to this.

He: We have the red shift which shows that an expansion is taking place. How do you get it into the pancake?

Me: I am not of this scientific world, so I find it easy to distinguish between observation and interpretation, especially when it has been accepted by you. And within that framework, funding is given because it's seen as cutting edge science and that's where everyone wants to be. Do you have time for us to go back in time? Uh, historically, I mean.

He: Yes, of course. I guess you knew this is a day when I work at home.

I didn't answer that. Should I say no and explain that time knew? Because he needed to understand how consciousness (C) is part of the formula for that. I continued on the breakfast menu instead.

Me: Gödel, who put forward the model of a rotating universe got a cool reception. I would say that Edwin Hubble's discovery of galaxies beyond ours and that they are perceived as moving faster the further away they are was an interpretation. It was the simplest interpretation at the time, but was not the only interpretation. Well, let's see now. I borrowed a pen and grabbed one of the paper napkins I had tucked away.

Me: You'll have to excuse my rustiness. The usual, the constant of light, the effect of wavelength on speed... or... the enormous mass of the universe.... Expansion could be replaced by something like this, gravity. Variations in Hubble's constant are due to gravitational variations and variations in the motion of the universe, simple. I love it when it gets simple. You fill in the rest.

He was silent as he took over the paper napkin and eyed it. I asked if I could serve some more breakfast cosmology from a traveler. He lit up again.

Me: Gödel's proposal based on Einstein's general theory of relativity meant that spacetime could be curved, causing not only return in space, but also return in time through its closed spacetime curves. So which weighs more now, me bringing breakfast from the future or the red shift?

We laughed and exchanged words about both the pancake model and that he would eat the one that was left while it was left and before it was made... Time travel humor always relieved the intellect of its pretense of being grounded in logic.

He: If I recall correctly from my rusty cosmology history, Gödel's universe would fall on not rotating.

Me: It fell on a lack of ability at the time to produce measurements to show that he wasn't right. You are there now but more basis is still needed to interpret the results. We need to understand how the whole universe is electric and how interacting electric and magnetic fields are of far greater importance in shaping the cosmic fabric. They are also the reason why we have an even distribution across the universe and why large parts of the universe move relative to each other. The Big Bang model does not hold up in the long run and both dark matter and dark energy are expressions of the need to compensate for the deviations.

I sketched some more. I had brought plenty of paper napkins to our private breakfast seminar. And a good pencil that I had used before. I continued.

Me: And... the total sum of these movements produces a dynamo and with it a slowly rotating universe.

We discussed the dynamo and Hannes Alfvén, the founder of magnetohydrodynamics, that he would come up in the future cosmology. I think he was a man who was not fully understood.

Me: The electric and magnetic plasma currents form a kind of tube through space. And they are far more important for how matter forms in space than gravity is. They both give off their own motion and hold themselves together with it. This gives them an incredibly intricate and beautiful dance. Unfortunately, it also makes them fragile. The time wars that have raged for hundreds of millions of years have caused damage. At certain focal points, a disintegrated galaxy can cause an entire stream to break apart. Think of it as cutting an electrical conductor in space and the whole circuit falls apart and it can get so bad that the event causes spacetime to collapse locally.

He: You're mentioning things you haven't mentioned before. I mean the time wars.

Me: Well, I couldn't gain your trust if I mentioned that time travel is not only dangerous and guarded, but that there has also been a war between parallel times. I could be anyone then. He nodded and agreed. I got to hear a bit about his rather long road to accept me. It touched me. I knew from experience but hearing it moved me. I continued.

Me: You don't have to be afraid of the thought of the wars. They take place over an incomprehensible period of time. Moreover, they are far above us little ones in creation, time wars are for beings who are like gods in their power and perspective. But let me continue... with the knowledge of the universe can be made weapons as well as vessels. Look here how you could have made yet another technological thing from the future along with your first exhibition project.

I had his attention again and this time asked for a pad. I sketched.

Me: Through coaxial plasma streams, you can also create a penetrating weapon that no matter can withstand. It is possible to achieve speeds higher than light with this. You use the same high voltage source and magnet, but you arrange it as a series resonance between coil and capacitor. The capacitor has to be flat, as I draw here, and close to the coil. Then one more trick. The distances between the plates must also be in resonance with the oscillation of the electrons. I draw it like this and the resulting motion focuses itself and forms an ion beam faster than the light straight ahead. Yes, with that setup, you could shoot a plasma beam through the ceiling and, at worst, hit the moon if it was in the way. This is a variant of Birkeland's plasma currents (Field-Aligned Currents).

He was intrigued by what I was saying so I ventured to elaborate.

Me: Space has no resistance to ionic currents, or more correctly plasma currents, so it is possible to achieve a form of superconductivity. I know about them because they can throw one forcefully in time, you see, they generate strong curvature. I sketch briefly here.

I brought the vectors down and compared the currents to eddy currents that occur in electromagnets and that cause heat losses and eventually strong folding of time.

Me: This forms a dynamo and if you build it with double bifilar winding like this you will get so-called free energy. Power, measured in Joules per second, depends on time. So with tidal effects in this curve, you can get more energy out than you put in.

I quickly checked with him to see if I could get more out... He nodded, which told me he was up for the challenge at least.

Me: If you understand these simple principles, you can create the energy, which outwardly is perceived as not consuming the necessary energy. It is the local time effects that make a craft built on the same principles not perceive momentum. In this way, it can also be said that energy is lost just as easily. Heal time and you have no energy problems or energy needs. This is what makes Kardashev's scale of how advanced civilizations use incredible amounts of energy wrong. It also shows that there isn't a Dyson sphere either that captures the energy of an entire sun, it's just a dense cloud of dust close in front or in a distant galaxy behind the star.

He was amazed and he shone. He is so great in that way. He gives of himself, doesn't play, has nothing to lose, only to gain. I mentioned a little how difficult it was as a teenager to manage to tune these five variables including the one for the capacitor disk spacing. He wondered if I had built it and succeeded. Well, built it I had and had gotten the expected measurable effects, but the calibration was too difficult and I only got the weakest results, but still. Principle partially tested. We laughed at how it was best that way, as I avoided visits at a premature age. I returned to the cosmic unifying forces.

Me: Neither dark matter nor dark energy is needed to explain the flaws in the Standard Cosmological Model. It's funny how such infantile emergency solutions of a flawed cosmology are given all the space in publications. Despite the fact that the sun is a pure alternating current generator of plasma and ions, no consideration is given to calculating these enormous electric and magnetic forces in the universe. They completely give the universe its bubbles between the electrical conductors with Birkeland's currents along the galaxies in the cosmic fabric.

He looked clever. Cute. And it suited him, he who never hid his face.

He: But you haven't explained the evidence for a rotating universe. How you could measure it.

Me: You have them everywhere, with today's instruments, but the difficulty is to interpret the results.

I paused for a moment to sort out what things I could bring up and in what order. In what order was I here, and when was the research published?

Could I anticipate that? I just had to go for it now that I was asked. I jumped right in:

Me: The Cosmic Microwave Background (CMB) is a good place to start. The distribution of temperature differences is small but clearly shows that there is rotation in the universe, which I think would be obvious given how the whole cosmos moves in swirling tubes and galaxies.
I find it comical that they chose the coincidence that this plane of rotation coincides with the Earth's orbit for the Axis of Evil. But the fact that galaxies have been found to have a 1% preponderance of rotation in a particular direction has not been called the Rotation of Evil. There is a one in a billion chance that it is a coincidence! However, it is not easy to get a coherent picture, but it will be possible. Once the fibers of galaxies are understood, which hold the universe together with electric and magnetic fields, an overall model using an advanced computer model will be generated of how the universe could move as a whole. Before then, several interpretations will be put forward, but the universe is rotating and it is possible to measure this.

We had some fun with the Axis of Evil and talked about the conventionalism of science and that Copernicus did not establish that the Earth was not the center of the universe forever. He only showed that the Earth revolves around our sun. One dogma was exchanged for another as a kind of unconscious pledge against reintroducing God again. I couldn't avoid sidetracking about the axis of the universe and what I knew about it.

Me: This is an area I am not allowed to fully enter, but I have said that I will answer everything I can that does not concern direct events. It was only when I left our spacetime here that I realized that time on all other inhabited planets was quite different. I know, you're looking inquiringly and you know I'm getting there. (We smiled.) They had time that more persists, versus ours that changes or measures change. The best way I can explain it is that time and events are not synonymous with each other. Time is relative and varies in the universe, generally it is far faster than in this backwater.

He: How big is the difference? I mean, is it a few percent of variation that could arise from relativistic effects?

Me: No, the differences are huge, this is a real backwater of the cosmos, we are like a living archaeological relic from the origin. I generally reckon that one day here is equivalent to two years there.

He dropped his jaw, as I expected, and questions followed, of course. I told him that we happened to be in the axis of the universe and it was not a black hole, but like a black hole we were looking out of. I gave him some substance.

Me: To get to Earth from the outside, you have to make a 90-degree turn in time. It's relatively easy to understand by how time and space change places, just like a black hole (I was sketching on paper at the time). But what happens to time there is unknown and it gives these incomprehensible differences between times, not only in speed, but in the way processes behave. You get rid of friction, wear and tear and get a much longer lifetime, not because of the speed but because the properties of time are different inside and outside the axis.

I sketched down some new relativistic formula comparisons for him to go deeper into himself. Otherwise I just had statements pulled out of thin air. He got that pondering expression when he caught himself doing some inner investigation. I liked it because then I knew he was taking me seriously even though it was incomplete, I mean, it was a breakfast not a dissertation. After he had sunk into the opening I had left him for a while, he sought my gaze again. Was I right that he would now come to the shape of the universe after its axis settled into place in his newly formed universe. I was right, he skipped all my encounters and moved on to the exciting cosmological questions.

He: This is inspiring and exciting. How far out have you been and can you say anything about the limit that does or does not exist for the universe.

Me: Absolutely! I have been to the limit of the universe and beyond.
I wasn't prepared for that, he coughed. We

pretended it was the morning juice slipping down the wrong throat.

Me: I have too little comparable data for a simple comparison, but I have times to different parts. The universe is far larger than anyone in cosmological science can imagine. I need to find a speed that I can measure with and that is consciousness. The speed of light (small c) is like measuring with a snail. Consciousness (big C) is instantaneous, a thought passes through the universe faster than it registers in a brain. But it is also not convenient. What is useful is the second fastest speed of the universe, namely how a thought travels through the matrix, via being after being, you might say, well with some hesitation. Let's skip what the matrix is, because that requires a deeper explanation of C. The times I have measured this speed, let's call it the speed of thought in the matrix, before it reaches the boundary of the universe is 11 minutes and 20 seconds.

Finally, I got a lovely reaction from him. I couldn't read him all the time and stay focused at the same time.

He: How did you measure this? Did you repeat the measurement and...

Me: Yes, it was with measurement deviations of plus or minus two seconds. You know, the only way to clock such a high speed is to use the instantaneous speed and that was my C following the wave until it stopped abruptly. The deviation of plus or minus 2 seconds was my variation in reaction time.

He: Really! This sounds like the universe has an absolute wall, and does not ebb into infinity.

Me: Absolutely correct. It is an enclosing wall! However, don't ask me to describe it, perhaps I can include it in the discussion of C when I visit you again. There is also no risk of anyone reaching this limit of the materialized universe, the one made up of stars and galaxies ends long before, after 2 minutes and 35 seconds if I remember correctly. And the visible universe, I don't really know when we passed, it was so early, just some percent of the whole universe. You'll have to excuse me, but I'll have to check these figures if I'm going to say anything.

He: This is astonishing! If I can believe you... I realize that, but the thoughts you're raising are mind-boggling. This means that the universe has an outer shape!

Me: Quite right and it's a bit like a flattened ball, almost like a Mentos, but not quite. It gets that from its internal rotation.

He laughed again and we had fun with everything I told him.

He: Here I am, with not only a time traveler, but one who can be called both a stellar being and beyond a stellar being!

We always needed our humor moments. They kind of shook the perspectives in place, not just our breakfast bellies.

He: Finally... What does it look like or what is it like outside our giant universe?

Me: Need I say hold on? (We smiled.) I haven't described how something can exist without time, but that's what it does. Events exist but not time. Without time there is no distance and yet there are constellations. Haha, I can see you thinking... Are there multiple universes?

He nodded smiling.

Me: Yes, many, incredibly many. All on a surface between The Two ultimate C's, like on an ocean, or rather in meeting between The Two confirming observers. And... If you don't offer me an extra tea, I need to leave this time for a refill. Nah, joking aside! Let's just stay in this and chat about normal things. If that's possible!

And so we laughed again. He put on some tea water and after a little stretching break we talked about ordinary things that were on the menu for this breakfast trip, such as the alternative values of the magnetic and electric constants in other universes and how they related to the Fine Structure Constant. Most of it I knew nothing about, but that way I also got to do some thinking and listen to his thoughts, with little interjections from me. The sofa was much more comfortable to travel reclined in, than the chairs in the kitchen.

We got into how the universe consisted of bubbles, what was called BAO (Baryon Acoustic Oscillations), which is a collection of masses of galaxies in the center and a wave ring around. Without explaining, I told him that it is the result of the universe not coming into being at one point, but across the whole at the same time as after a sudden rain of drops. In the Big Bang model, there is no explanation for the universe being flat, so it doesn't make sense. But an emergence over the entire surface makes a flat universe obvious. So do the constant gravitational waves of the universe. But how, would be a consequence of big C, which I have not gone into, and which formed the core of the God formula.

But I have to say that what tasted best as dessert was to tell my hilarious mistakes. It built trust! He really did see me as a friend! I breathed it in and sighed it out slowly, as if relaxing in the warmth of friendship, which I did but also more. My universe, which I bore within me, expanded slightly to return gathered around the heart in me which we shared. We were silent together before the unspoken greatness. And remained so.
When he opened his eyes, I had taken another breath, as the universe does. Left behind, swaying gently, were only rings from one drop of the universe, reminiscent of a Japanese garden.

Part II – Echoes

Had it not been for the fact that the remnants of his tea time with me were still there (but without the paper napkin), he would have thought that his busy thoughts had passed him by and that he had sunk into the morning's inspiring work.

Discontinuity in time ebbs away until only the pristine surface of thought remains.

The remnants disappeared after a while, but I was still there, he wrote me into his time. Just a few lines, with no mention of me, but enough.

Our laughter would last. What is more beautiful immortal than that.

From our after hours on the sofa, thoughts
flew freely.
They are still flying.

They are just waiting for someone to land them
in their hand.

How do you know that a scientific dogma
has been thrown overboard?

Well, when an observation is named
the Axis of Evil.

Doesn't that say it all, that bringing
God back into science
threatens science as the supreme God?

How did I know that the Earth and the solar system were at the center of the universe?

First of all, everyone knows that.
Second, there is only one place in the universe where you have to make a ninety-degree turn in space-time to get to it.

A ninety-degree turn in spacetime is kind of a black hole,
not really, but more of an axis.

Paradoxes are never paradoxes.
Paradoxes only say that there is something paradoxically wrong
in our constructed conception of reality.

Reduce the statement of Copernicus that the Earth was not the center of the solar system to a statement of fact,
not a *new* dogma.

To state that the cosmic microwave background is in the same plane as the solar system is far beyond chance, not a threat.
Paradox precedes paradigm.

Is Earth time different?
If it is!

This crazy place has time that changes.
It makes you think that time is a measure
of change.

Time is.
Time is consciousness bearing.

That's when you realize that *you* are a vortex
in the ocean.

A rotating universe is no more difficult to understand
and measure than the Earth's rotation and its cause
of swirling hurricanes.

Look for the cosmic Coriolis effect.

If the Earth is again placed at the center of the universe
the prevailing cosmological dogma falls.

Big Bang becomes *Silence, No Bang*

No cosmic catastrophe,
only personal collapse of prestige.

Parallel planes of existence act perceptibly
and
invisibly in all galaxies.
Electric and magnetic fields interact visibly
and invisible between galaxies.

Just because they cannot be observed does
not mean
they are not there.

Dark Matter does not exist, only the name
itself misleads.

Go back to the observation
before the interpretation of the observation,
and before the name that gave the
dogmatization.

What you are left with is cohesive force that is
not gravity and is far stronger than gravity.

You find that in the electric universe.

Dark energy does not exist.

For those who are enlightened and not
trapped in the spatial realm of imagination,
the answer is not in darkness.

Without the Big Bang misinterpretation
there are no inaccuracies.
No need for invisible and contrived
answers either.

How do I see invisible matter?

I phase into a parallel plane of existence
with its value on my Planck's constant.

How do I pass through a solid wall?

A wall is never solid
It is only composed of waves
interacting with each other.

With a constant that does not interact
the wall is no more solid than waves
that just pass me by unnoticed.

What forces?

I bend space with thought and have the
force
you cannot understand.

Consciousness is the base of time and
time of space.

My thought is the movement of my
consciousness.

Force is what you think you're measuring
when I let my consciousness take on
a new direction
of space that's been forced.

It is normal among the life forms
of the universe.
On Earth, it is paranormal.

Galaxies so old that they could not have
existed
so soon after the Big Bang?
Then we extend the age of the universe.

We laughed.

Then we keep the Big Bang twice as long
until a new telescope sees more,
too far away.

Good tea this!

Time is closed.
This means that the universe
exists before it is created.

It will be fun to be part of
the beginning.
Not long left now.

It is through the existence of
the universe
before it comes into being
that it can come into being.

It was the movement of creation
that
gave rise to creation.

Sure, but don't tell!

The Two Observers
have not yet risen from the sea.

They seek each other in the creation
they gave birth to.

What happens then?

Apart from a beginning in
circular time?

We get peace and all the cosmic wars
stop.

All cosmic wars
have been about wars for existence.

The one who got the most confirmers
became the greatest God.

Does the universe have a purpose?

Absolutely not!
No, it's much bigger than that.

The universe is the book in which
The Two find each other
after they met.

Something was wrong. I needed to change my beliefs when I realized that extraterrestrial seasons were synchronized with Earth's. That's not possible unless we, our solar system, are central to all parallel times.

Believe me, the path of my thoughts has been long and arduous. I have a stubborn mind that likes old understandable beliefs. It makes sense, because who doesn't?

The Axis of Evil.
They didn't have to call the observation that.

The pancake pan was a pan for Pan Pizza
But then you need the pizza maker who rotates the dough to perfect roundness and flatness.

We laughed. What a time it will take to overcome the Copernican worldview and the dogma it became.

Without the Copernican dogma, it would have been called the Axis of God.

The temporal police

We talked about the innovations of the future. Why not bring them here. He was excited about all the applications I told him about, and which were easy to adapt.

It would be possible if I reinvented my inventions again, but in the past tense. And so we reached the inevitable topic of temporal police. A heavy topic for me.

Me: I thought it would be appreciated if I left some innovations here in this time that would revolutionize technology and help save the Earth. But that was before I realized this police existed, which, well, can detect time distortion.

He: Now you're making me curious. Tell me more. How do they act? How do they detect?

Me: Hm, well, it's more complicated than I want to get into. I really want to avoid that, even thoughts can attract them.

He: Now you sound a little sci-fi and a little paranoid. I'm just saying. Go on, please.

Me: Future beings, you would call them aliens, are anxious not to have their past changed by other Earth visitors. They may not come to the backwaters of Earth to manipulate time, but they can and do make changes in the parallel time continuum like multiple rivers affecting each other's flow. This is one of the causes of the time wars that have raged for ages. I have written two books on the subject about why and how.

He: Really? Where can I find them?

Me: Uh let's see. Uh... yes, they are published, you can find them in your time. But not in English, it will take around three decades before that happens?

He: Why the delay? Is there anything I can do?

Me: Well... First of all, thank you for offering me help. But I'll explain why, with the temporal police. There are two kinds of police. The police from the future who do not want time travel to be known. It's complicated, as it is with all the parallel futures that affect their timeline. They usually work as discreetly as possible. They influence elections. It is not seen as manipulative. I protested and wanted to expose this and make humans aware to let them choose their future themselves.

This got me stuck in the time wars. Being the most primitive on our time scale, I had an advantage and that was being located in their past. This meant that I contacted them all. And, yes, so I became a dimensional diplomat for peace.

By drawing attention to the time wars and the

time warriors to this time, I was a threat for decades. Of course, they kept my books from reaching people. Only a couple hundred were delivered to readers, but I managed to make a difference anyway. By 2020, they had all surrendered and were no longer preventing my books. In fact, they, the futures, had begun to inspire new contactees to research UFOs, as time travelers. Knowledge of time was no longer a kept secret. That is how I know that I have somehow won the time wars because peace was brought to the universe.

He looked at me. It wasn't his normal look, and I couldn't read it. Had I just said too much? I was used to so many moments being ruined when I reached forbidden subjects. A sudden and total denial can occur. I called the phenomenon caused by reality intrusion. I waited, took a quiet breath and hoped. Finally, he filled the void with a question.

He: But this hasn't happened yet...

Me: I'm sorry, I apologize, I have some time confusion. It happens often, too often. My timeline is linear for me but compared to yours it has made loops.

I paraphrase what I said. In your near future, here, the time travelers have been convinced that the only way for all of us is to cooperate, and that includes humans. This is the only way to stabilize time so that a time collapse can be avoided.

He: Time collapse? What do you mean by that? Curious again.

Me: Uh, this topic is difficult because it has so many branches. Or actually, time has created many loops and that makes it difficult to explain without visiting all the branches needed to cover this topic. I promise to return to the time collapse when I answer your question about how I became a time traveler. It is not forgotten.

So, the future has changed, or more correctly, in your time, it soon will. That makes this category of police no longer a threat, but the Earth-based temporal police is worse.

He looked at me with curiosity again. But did not smile, yet. It's better that I catch his attention with a more playful post.

Me: If it weren't for this temporal police, I would have met you earlier and you would have built an even more impressive science project.

I smiled. He laughed. I had his playful attention again.

Me: I'll tell you how, but let me finish about the temporal police. These are run by mainly top secret agencies in your country. They have access to unfamiliar technology so they can monitor every major entry into your time. They are the truth behind what is called Men in Black and also black helicopters. They operate totally ruthlessly and have killed several of my colleagues. I promise you, true inventors are like you, they are driven by curiosity and work day and night. They don't leave sudden notes about why they don't want to live anymore. It is impossible! And even to commit suicide in a similar way. That's why I can't, uh, couldn't visit your country anymore. I've already been exposed.

He: That was a heavy statement. It's a lot to take in. I mean without much evidence, yes, tons of knowledge. Insider knowledge too... and you have to realize that this is a bit much.

Me: Yes, I know it is. But let's deal with some more. I'll give you the evidence you could have built for your science project.

I smiled at him and he lit up. Now for the fun part. To inspire with practical application.

Me: Well, to create anti-gravity, you don't *create* anti-gravity, you don't even cancel it. You create an object that is isolated from our time and that means that gravity and moment of inertia come with it. Let's start with project one. You need a very strong electromagnet and a high voltage source...

He lit up when he realized that he had built just that for his advanced exhibition project. His eyes told me that he was wondering if I had visited him at the exhibition? I had deliberately shown with my hands in the air as I spoke, the size and shape of the objects needed.

Me: Add to this a huge round insulator, like those that hang from high voltage cables. You know, the ones that look like bell-shaped UFOs.

We chuckled together at the thought.

The high voltage is applied across the top and bottom of the bell disk. The insulating disk forces the electrons to travel over a large surface area. And now we have your huge electromagnet inside the insulator that makes the electron rotate around the surface.

What are we missing? Oh yes, the lowering of air resistance to make the electrons spin around the insulator. There are two ways to do that. A container with reduced air pressure or near vacuum. You see, you had all the equipment!

He shone like the child he once was. He still is, by the way. If he wasn't, I wouldn't be talking to him.

Me: This would create an upward motion, and to go down you simply reverse the polarity of the high voltage. To make it controllable, you add three half-spheres for the high voltage underneath, instead of one. This tilts the rotation a bit and you can move the craft sideways.

Something else, a bit more advanced but no problem for your age then, would be to add a magnetron to accelerate the electrons. A bit dangerous but you could have flown and maneuvered it outside via suspension cables.

He laughed and we talked wildly. I made some sketches on paper and put in the scalars and vectors and the overall resulting motion of the sum of the electron's three-dimensional spin. I made some brief notes on how this curved space goes around the insulator. He would remember the sketches and notes, but not where they had disappeared to.

He gazed silently for a moment and suddenly burst out.

He: But this solves everything with the Tipler cylinder! You don't need to make an infinitely long cylinder to produce negative energy!

Me: Yes, having rotating matter like this is so simple that even a teenager could have put it together for a science fair.

He: I'm shocked. I am amazed! This is incredible! I don't know what to say. You gave me the proof! And not only that! Applicability. I love it. With this, one can immediately test the undeniable validity of the equations.

Me: You see, I had to choose between you then and you now.

He looked at me and kind of knew what was coming.

Me: I chose you now. For several reasons, but perhaps the most important one is that you got to live a long productive life. Not a short revolution where you would have been oppressed like a fraud. I could not do that to you.

I could not hold back the tears. We fell silent. I think he had a flashback of me. That unknown man standing in the background, like some distant but proud relative who couldn't leave, not until he just had to.

But what he couldn't understand at all was that I still had this journey to make. Standing there and staying in place, for me, had not yet come.

I know he felt deeply affected, unable to understand why. I guess he was affected by my tears. And the images in his mind are just coming to life. I couldn't tell him that it was his proximity to this upcoming event on my timeline he was remembering in reverse. For me, it was my next memories as I stepped out the door.

I don't know whether to curse
or bless moments that reconnect.

A heartfelt desire.
A heartfelt pain.

Two moments in time
cannot be closer together

than humans are.

Intertwined life
intertwines time.

Unconscious humans
still intertwine time with consciousness

and so desire to be controlled.

The strongest feeling
is the closest moment in time,

to the event in time.

Emotions bend time.

We call it time of pain.

You do not rise above this reality.

You learn emotions.
You live emotions.
To the fullest.

Field E and field H are companions.
Together they form the total EM field.
Where one is, the other is.

Movement as scalars
in three dimensions,
results in motion
of the fourth dimension.

To travel in time
you need to initiate an E-motion.

Yes, emotions cause time distortion.

Emotions are the concept of movement
in time.
Over time, one learns to live
emotions fluidly,

in all directions.

Without a machine
to carry you,

this is the life
for a portal walker.

I just made an entrance in time
as I do when two times are close.

Nobody noticed it.
I am a natural part
of the event in time.
I just enter their consciousness.

I looked at him from a distance
with the fresh memory
of us recently sitting down.

He could feel it
far across the room.
The emotion that has no words,
only an intense connection
that set him in motion
so that one day we would meet.

I will always remember him asking:
"Have we met before?"

And I get just as filled
every time
with emotions
that tie time strings.

Surviving a time shock

Not many Americans have good tea in their homes. The coffee, well it's not what I call coffee. It is usually so thin that I sometimes stop and wonder if I got tea anyway.

But some people brought other habits with them to America and I am so glad he did. Although I would probably say given his age that I am glad he kept them. He read me and I read him. I could sort of tell in the jasmine scent that he was on to something, and he was.

He: There must be permitted time travel, right?

He had a funny way of asking almost philosophical questions. He bent his head down and looked at me a little under the bangs and over the glasses he wasn't wearing.

I gave him the same look but more out of the corner of my eye as if I didn't want to turn away from my thoughts which he was wondering about. And he got a smile from me, quite similar to his smile. It made us both chuckle because the signal for tacit understanding had been given. Yep! Now it was just a matter of exposing oneself.

Me: Yes, there are, but that doesn't mean they're any less dangerous.

He was quick to ask to fill his worldview to a fuller picture.

He: Is all time travel dangerous?

Me: They don't have to be but can be. The simplest way I can put it is that natural time travel, yes it exists, has built-in safety mechanisms that prevent the collapse of reality.

I inserted a little passage and explained a little faster because this was stuff he already understood.

Me: If the Universe as a creation were to contain laws that did not protect against the crossing and looping of time... or that time could reverse suddenly or worst case scenario where times open up uninhibitedly to each other, well then the universe would collapse. It is just that without time, i.e. with collapsed time, the universe would not even exist. The beginning would have dissolved. A logical consequence is that only the universe with functioning and self-protecting natural laws remains and is born. The others disappear and never existed and thus never even came into being. Fine-tuning comes with a functioning universe. So there you have it.

I continued at a more calmer pace, fitting the scent and mood of the question. We let our cups follow.

Me: The safest time travel is with the flow of time!

We laughed together because we were thinking the same thing, both he and I along with our tea were traveling safely through time. We sipped it. I continued.

Me: There is more than us and our tea (smiled). There are travels in time that do not mess up or fully obstruct. These are related to degrees of portal entry I have mentioned and referred to as KRS. But let me give you examples so you understand. They often go under the concept of visions. That is, you see and even stand in an event, a disaster for example, but without interacting. It's not a vision, it's a time connection. And it is certainly not an illusion or a psychotic episode. It is a natural passage of time because time is not a line, but a fabric.

He: Now you have me trapped, he said smiling broadly. If you can make sense of this all the way through, with concrete examples, you've made my day. I've heard your words but you know, I want to know more concrete things.

We smiled together and took an agreed micro-pause with a mouthful of tea. The calm before the revelation showed that we would both get there before our cups were finished.

Me: I'll get there, I promise, but we need some casual interactions too as time heals together despite the passage having occurred.

There are situations where people have crossed

into another time, but time always closes in on you if you do. Also, your memory changes.

I know it may sound like time is intelligent and it will be seen as a pseudo-scientific explanation. But you will find out more about it.

But, let's go to my current example that gives you the answer you are looking for and more than you asked for, reverse cause-effect. What I describe here happened after what I call the Arrival, so it hasn't happened yet from your perspective.

Call him Ben. He is an established scientist who has become known for his unique research and unique analysis. He is an American with a professorship like you and originates from another country. That is all I can say here at this time. Perhaps I can bring you together after what I tell you has happened in your timeline's position.

I traveled directly to him for a specific purpose, as I also did to you, and the moment I shook his hand I got a shock, it was a time shock. I know that it can happen. I know it usually happens, but prepared I can never be. I just have to trust that it freezes time for both of us in the meantime and that my apology is taken as credible. This is usually not a problem as I have learned to keep smiling and looking into the eyes even though inside my mind is shaking all over with aftershocks.

On the outside, in my mind, it looks like a vision

of us having met before to prepare for a meeting we will have when I shake his hand.

He, on the other hand, will take it that he understands that everything I am telling him has already been realized. But I haven't had that meeting yet, I've been thrown into it right after.

I met him when he was growing up. And because for me it happens after meeting him in his future in America, every word comes out of my mouth as written. They are so obvious that I cannot and will not deviate from them. I was just there in his future. Call it a sense of destiny that you don't want to destroy because it is beautiful. Call it that time has a will of its own, and it does, but it's more like a magnet, a wave that gets stronger and stronger because it resonates with itself.

Because the event of his childhood didn't happen until he met me, he won't remember it. A memory will come back, like a dream. Something he dreamed, but which is still real. He will have a sense of déjà vu about our meeting.

Only if he consciously starts to penetrate his own mind will memories of his “new past” come up. They don't quite fit in, but after a while his mind connects it organically. The brain has the ability to assimilate the appearances of different times into a timeline. In fact, it's one of the laws of nature that must exist for the universe not to collapse

In fact, it is better for reality if time collapse is

contained and not spread.

I looked at him and knew he had stopped, but not because what I was saying wasn't getting through, but because he was putting it in place. I continued.

Me: An encapsulation of a time collapse is also an encapsulation of a reality collapse. A collapse of reality is what psychosis is all about. There are more people than you realize who have violated the time laws of the mind and then collapsed themselves. That it happens together with mind-shattering events is only natural, because that is when the mind is outside its own built framework and can cross all previous boundaries. But keeping it together is something few people can do.

He savored and exclaimed a realization that had not yet fully landed.

He: This is amazing, just amazing.

I let him continue, fumbling a bit to put the pieces together, and as he said, time travel requires interdisciplinarity to be understood in its full breadth. I nodded him in response, most of the time. He felt inclined to write another book about what the future would bring, and that we wouldn't be able to do it on our own and didn't need to put our faith in our own miracles.

I got caught up in my nodding and added that this is precisely what the Arrival means without having to explain more at this point. I also avoided answering how we survive into the future despite being doomed to extinction. When he put down his cup, I told him the broad summary of permitted time travel.

Me: What I described here is advanced and permitted time travel. Manipulation is to first go in ahead to direct. This is entering before and amplifying desire. Amplified desire and the realization of someone's life brings amplified time. That is what time wants, because time and consciousness cannot be separated. I mention it but won't go into that now.

He: Maybe I'm imagining things, but I find it so much easier to understand you now. Have you made another trip to me that I don't know about?

I nodded, holding back emotion. He saw it, waited respectfully and let me continue in any direction. We were in complete harmony now and he would understand the last thing I wanted to add.

Me: I knew you were on to me, in time. All important encounters in time are preceded by a pre-event, after. Déjà vu is recognizing before even though it is after. After creates before. Now you heard me say retro-causality with new words.

That you understand me more easily today is due to synchronicity. Time travel gives rise to synchronicity. It's not something the mind invents to find meaning in life. No, it's the meaning in time that the mind suddenly sees and thus lands in and realizes. And I should probably add what you noticed in our dialogue and synchronized tea drinking: two people with two events that are in sync, equals telepathy.

This time his “amazing” landed quietly. And I respected him with the same serenity. We were a sea and a pond in a Japanese garden. Two lotuses at its surface and deep below, unseen, we shared roots in the bottom of the pond.

We sat there for a long time even though he had lost the meditation of his culture. Time had no limit.

After a full hour of silence, we got up and the clock had moved 15 minutes.

Reverse causality makes the fabric
of creation.

The fabric of time keeps reality
from collapsing.

The fabric of time must be kept intact
at all costs.

Time wars are the worst crime
that you can commit.

Telepathy is nodes in the waves of time.

Regardless of distance in space or time
they both oscillate as one
in the moment.

Nature has laws against using shortcuts.

Permanent damage is the price
your soul pays.

Time has always been a fabric.

Your perception has been limited
to a thread in one direction.

Reality is just a surface.

When you add time, it gains depth,
the depth you call space.

When you remove time

reality is a surface again.
That's how you take a step through a
portal.

A step in time
is no further than in your mind.

Lose time for a moment
and you took the step.

Positioning my superposition

I loved his classes. I'd been in all of them, just not present like today. I always focused on the same chair, but usually let my focus flow out across the room. It gave me another fuller absorption, one that my body could carry back with it.

Today, however, the purpose was different. I was to write history without leaving a trace, and that was a challenge in itself. But the big challenge was to describe time travel in a way I could. In a holistic way, and then most easily with motion signs with a mathematical connotation. I would not be able to arrive at the God Equation through Earthly math. It was too complex for me, who was not used to painting correlations in that way on a daily basis.

I could certainly figure out how to divide two equal numbers and get the result 2, but the huge detour... It made me deeply hesitant. Besides, I only had 17 hours before the next class was due to start in an hour. I needed an unaccounted margin to clear the room as well. And the stomachs would reveal that our time was longer than it is supposed to be.

The problem of the stomachs giving away the real time was solved with an improvised lunch from Subway. But I couldn't get them beforehand, as I needed to, because the opening would come only with our successful meeting. The waves of these would give me an increased likelihood of interaction with the matter of this time after our meeting. A good side effect of that would be that our paper-wrapped lunches would be up-to-the-minute fresh. I would just have to keep my backpack a little discreet so it wouldn't show that it was empty, before we got started and created the necessary waves of our meeting to pick up the meals from Subway afterwards.

I now just needed to create the perfect conditions for our meeting and it was just a matter of timing. A point in time is like an announcement being made at the right place on the scale. I needed to get up just right so that I didn't collide with the other students. And if I catch his eye and his interest, a tunnel is created past him so that no one stops and starts talking to him or asks any follow-up questions. It worked. He lit up, the students were diverted and directed their tunnels of thought to the hallway outside and the route to their next destination. I smiled, thinking about how close they are to jumping to the goal instead of walking when a tunnel of thought is created in their introversion.

Would they realize they jumped in time when they jumped in space? No, they would only vaguely remember the path, as if they were walking in their own thoughts, which they were. Their trapped thoughts with the goal in mind were just what was needed to isolate their consciousness from the time and space around them. Such a thought would look like a tunnel, if one observed it. I smiled again at how artists have the ability to capture realities, even what it looks like between realities. The movie Donnie Darko (2001), was one such work that was so spot on.

I watched him as I walked down the steps to the desk, and with each step the exit got closer to being in the room. His thought was captured. I knew I would get through to him. It doesn't always happen and yet I have to make such "attempts". They are attempts and they are also wave makers that eventually bring out the event that creates the concentric waves. Attempt upon attempt, synchronized rhythmically, creates the standing wave in space-time which in turn creates the event. It is my secret as a time traveler to be stubborn, to do and do again, again and again. Until the thing that is supposed to happen has happened. Have I created the event? Yes, I have.

Does this make me the creator? No, but a conscious and careful writer of my time in interaction with the other strands of time. Sure, I could force it, but others would perceive that as an intrusion, i.e. deliberate sudden impact on their time strands. It is an act of war and comes back as a retaliation. It is easier to accidentally start a new time war than to create cooperation.

My mistakes would only form a small intrusion, but could be seen as a deliberate attack by a time master. Such consciousnesses have for humans godlike qualities and abilities. It is out of thought and time that they shape and reshape planets. I write my own time consciously.

My spatial steps strengthened my steps in thought. We were connected. I was in his thought and the others had not even seen me. They had only perceived without seeing or realizing that his tunnel in time was for me as a visitor. And he saw me because he was in my reality which he perceived as his. For he who knows, thought has no limit to what it can reach and thus no speed. The little c of the speed of light would today be replaced by the big C of consciousness. It excited me too, but mostly because I reached there over an abyss in between. That's how it is when thoughts have not yet built the foundation of the reality we stand on. It was time to start. He began.

He: I saw you, I thought. But then I looked again and I was surprised, and sure. I was a little distracted by nodding goodbye.

He paused for a moment before continuing, He had found my time line.

He: I didn't see you come in here, but weren't you sitting in the seat where no one sits?

Me: Well, that's true! I always place myself there! From my smile, he thought I was joking. And he laughed, both at the joke and at the joy of seeing me again.

He: Did you know that it's always empty?

Now it was time for me to introduce to him that I travel in space-time and how I enter.

Me: You didn't see me come in because I didn't come in, not through the door anyway.

He stopped and dropped the books immediately, the ones he had just collected, to get ready to catch my thoughts instead.

He: Did you do what I think you did? Did you walk into this classroom's time?

Me: Yes I did! And I went to every class this semester to reach you.

He: Please tell me more! Let's sit down, because I guess this is going to take time. I'll just lock the door first...

Me: No need! We're not quite in your time, so we'll have plenty of time to ourselves. Yes, let's sit here. No need for pen and paper, we'll use the blackboard while we're here.

We started when we sat down, quite close, a sign of intellectual intimacy and deep interest. I knew this because without it I would not have been able to enter spacetime so clearly and accurately. I had leaned my small backpack against the chair at the side. It was now filled with our afternoon lunch. The wave of our meeting had reached Subway by morning. So I knew we would have many hours! But now it was the entry he wanted to know more about.

Me: I'll start, I'll describe an entry in time with this example. First of all, I have been superpositioned here every class this semester. To be here, I needed a closed room. I'm glad you didn't open a window! Because then I would have dissolved into nothingness!

Our humor synchronized us rhythmically. The laughter shook off the imaginary thoughts and freed our cores. The consensus that synchronizes our respective consciousnesses is unbeatable. I continued.

Me: You've noticed that chair, that no one has been sitting there during your classes this semester. Why is that? The answer lies in the fact that you noticed it. No one is sitting there. Have you ever noticed anything like that before.

He: No, I haven't. That's exactly why I remembered that chair.

Me: No, it wasn't. (A short pause for him to leave room in his thought.) You can figure this out faster than I can. Take the number of seats, the average number of students, the number of vacant seats, and the probability that a particular seat will be empty for the entire semester. Then include the probability that you observe that happening and that that particular seat is empty. Why don't you remember the other seats that were empty often, if not always?

He nodded slowly, did a rough estimate and understood what I meant. I continued when I saw that his eyes were clear again.

Me: Let's start from a different model. In a superposition, I would theoretically be all over the surface, in this case the classroom, with equal probability until someone observes me and determines my position. Recognizable, yes, but not complete. If the particle wave is consciousness, mine as an example, what will be my position to you and the room?

He chuckled, but I could tell from the sound that the thought hadn't quite sunk in. That is, it hadn't gotten real enough and seriously enough. I continued.

Me: To put it simply, probability waves are positions not fixed in time. And time in this case is your time. (I paused briefly.) So what about me, my position and my time? According to the superposition principle, I am located with my thought throughout space, call it the presence of consciousness. And now we come to what is not included in a single observer's time frame. I am consciousness that fixes me in space-time, just as I as an observer fix a particle wave through measurement. If I understand my thought as the wave it is, just as I am such a wave of thought, I cannot only superposition myself in space. I can gather my superpositioning into a seat in this room.

He laughed in surprise. He knew I was going to continue and sensed the continuation, and that's why he wanted to hear more. "Go on, go on", he said eagerly.

Me: I contract my superpositioning, in quantum steps, until I am fully present at a defined point in space. It is in fact also a phasing-in in time. It happens completely without light effects. It doesn't happen with light at all, but in consciousness. (I paused briefly.) Did you notice that the other students didn't see me? And when you saw me, they no longer saw you. Their behavior changed from that second of our connection. So I can enter a room without anyone noticing. To enter time is to enter consciousness. I will come back to that.

He: I am astonished. I... go on.

Me: I see you're losing your ideas about time travel. That's good, because everything you think you know is wrong, because it's limited by known knowledge, well, accepted beliefs. You and I were phased-in to each other's space time and when such resonance occurs, you immediately adapt to me. We formed what is known as a tunnel in time between us, although correctly it would be called a tunnel in the room. Let me jump back. If I center my superposition at one seat... how will the others react? Well, they will statistically significantly avoid my seat, as it is invisibly occupied by my temporal presence.

Now he was laughing, and so was I. He waved his hand for me to continue. He didn't want any words to interrupt.

Me: I am good at time derived from consciousness and I am good at the superposition of consciousness. This makes me a portal walker unlike others who are time travelers and need equipment to move in this multiverse that is forming. There are no parallel new universes emerging without strings touching, playing their sounds of time frequencies together and separating. So drop all current primitive multiverse theories. You can play with mathematical Lego, but the universe does not hold itself together that way even if you get a building together like that out of your mathematical Lego. Something I will explain with the God Equation.

Now it was impossible to hold back any longer. Lots of thoughts connected, ended in ends and asked questions. New thoughts answered and connected. All in full playfulness, just what I loved about him. I felt so full myself from all our chatter. Anyway, I managed to indicate that there was more.

Me: I want to highlight it from a side angle here. This presence has been experienced by humans throughout the ages and given different names, a spiritual being for example. Now what have I said? Well, that mediums, shamans and religious people of all times have touched on the parallel realities, which are parallel times and parallel space times. Their thought strengthened with faith has led them through the illusion of one reality, to the multiplicity. In most cases as light touch and they have then been called spiritual. In some cases tangibly and with frightening physical effects and light. In a blinding light it has been taken as a messenger of God, but without light sometimes fear took over and they were interpreted as demons. I choose to enter discreetly. We smiled in mutual understanding.

Me: So, what happened here in the classroom? You saw my consciousness! You saw me as another observer! Thanks to you noticing me, I became real. And you needed to look up again to confirm that I am here. You did, and I was here. The law of temporal events begins with an event, continues with the confirmation and is finally written into time by consolidation. Everything that happens three times continues to exist. The consolidation happened when you took my hand. So I am here. If you hadn't taken my hand, I would have been decimated to a wandering thought that you had every time I sat here and you looked at the visibly empty chair. Time is a fragile construct and events can easily be lost.

He looked dull and the only words I remember him saying were "horrified" and "delighted". I couldn't listen properly but needed to stay focused. Staying focused is like remembering what is to come or rather wants to come. From that, I can only do my best.

Me: So, can matter come from nothing? Yes and no. Can an observer create the outcome? Yes, absolutely. But what happens if two observers exist and they meet? Then what humans call absolute reality is created with matter, positions in space time and interaction.

I paused briefly. It's always difficult to find the right words when I can't confirm. I said it anyway.

Me: You've noticed that I've been avoiding your questions about string theory. That's because you have to put aside everything you've been working on and start from scratch. You have to start with the God Equation!

What does a laugh sound like coming from a throat that's out of breath? I heard it, barely audible.

Me: So, do you have time for me to show you negression on the blackboard?

He ignored my words and I guessed that he interpreted me as slipping my tongue.

He: I have less than an hour, so sure!

Me: I have seventeen! I suggest we take my time!

We laughed together, which I just love. I took the chalk and held it for seventeen hours, including the break for our Subway. He was still unaware that we were actually taking my time, and I didn't remind him of the clock during my entire time at the chalkboard. My consciousness, the big C, was speeding us up so the clock barely moved, just as I would come to show him on the board. The time's pace was set by my impact. A completely blank surface, white chalk. I love improvised symphony. I closed my eyes and thought, "Bear me!" and with a tremolo that said, "Please, all the way".

The God Formula

Nothing more stupid can strike me than when I start thinking about how the afterworld will look at what I do. It happened to me just now when I was thinking about how to present what I promised. How should I write this so that I can publish it understandably for a future audience that wants to avoid mathematical concepts? One focus was enough, to use my rusty mathematical skills to describe in an explanatory way what I meant. I need to reconcile them and get it together within my 17 hours.

I chose to use a mathematical notation reminiscent of the extraterrestrial one I learned 22 years later, but with corresponding Earthly characters. That's how I do it, I decided. I mention the calculation procedure but do not perform it! Then I avoided the problem of leaving unallowed traces. A calculated formula and matrix is a serious trace, but a mediated thought is not. A mediated thought becomes a track of thought in which the thought that belongs to its own time rolls more easily. I was only facilitating and it would not be seen as me tampering with another trajectory. This would not be noticed! It would also erase me to some extent, but I am used to that. I was lying to myself.

"Turkey sub, generous on flavor, low on fat." That's what the SUBWAY restaurant sign said.

"Amore Mio, our vegetarian SubMelt®. A light bread filled with semi-dried tomatoes, cream cheese, mozzarella, American cheese, red peppers, red onions and pesto dressing". I weighed and decided to exclude the Chicken Teriyaki Salad, it would interfere too much with the cutlery. Avocado Sub, well, not filling enough and double subs would raise the thought of time over stomach capacity.

It's funny that I'm standing here choosing, the day after, even though I know what we will eat together yesterday. It's not the mind playing tricks on me, but this is a property of time. Time gives me choices and that is part of the function of time. Did I manage to convey that yesterday? I don't think it went in consciously, but it's breathtaking that every encounter is choice by consciousness (big C). C wants us to connect. You could say that C knows what it wants because it seeks itself through time and yet it is the timeless. When C knows strongly it is experienced as destiny. I call it a strong current or deeper path of thought. I kept thinking back to what we said yesterday and what it meant here in the queue at SUBWAY.

How is it that one path is deeper than another? Perhaps the first question to ask is how is such a path experienced? A deepened path has more meaning than an unprepared path. Sure, you can go against such a path, but no one wants to because it is pre-selected. By whom? By you, you chose it because your big C knew where it was going. The time your C created gave you the discovery or rather the rediscovery. It is the inverse relationship too. The time that C created at the speed indicated by little c (the speed of light) made big C want to come into being. How is it possible to both exist and want to come into being. This is precisely the difference between time and non-time. In non-time I am not positioned in space because I am not positioned in time. In superposition I am everywhere, in all time, but for the one who has experienced everywhere it is not to exist.

To be met, with eyes, greeting hands, and the C of the soul illuminating the inside suddenly formed, there is nothing greater. It was out of the meeting that C was created. Actually, the meeting is worthy of being called a love rendezvous between two infinities. The Two superpositions that confirmed each other's existence and after that could never take their eyes off each other, 2C = 1, once we ignored the speed differences of c (speed of light) and C (consciousness), and allowed E (event) to become a singularity.

We had an encounter that made us detached from surrounding time. We were an event, E. To maintain it required wave-synchronization. Therefore, I kept the beat synchronized to our consensus. That way I could keep my time for both of us like synchronized vibrational clocks can give common time in two different places.

Me: So let's assume that everything is consciousness, big C. How do we get to the other fundamental concepts like space, speed of light, etc.? When C meets C, consciousness meets consciousness, the superposition collapses into reality. When the connection over time is measured, the second particle travels back in time. That time tunnel is consciousness C... So it takes two different superpositions to form reality, or rather one particle of matter (m) fixed in time. I pointed back to some formulas and reminded that this is true when C observes, thus we measure. So I have put consciousness C into the formula by knowing what C is or how C arises. I promise to come back to that. It's easy because the formula is only true through C. And as you can see, time disappears, which looks like the particle is traveling backwards in time and knows the outcome of the other particle.

We went on to talk about how quantum physics was stuck because it had failed to introduce consciousness mathematically. But that was not possible without understanding the origin of consciousness through E, the event. Speaking of E, we came to E as in energy.

He: Do you mean what I just saw... that force does not exist as force that we learned and built around, and then the conservation of energy and oh my, the conservation of information.

Me: Let's take one thing at a time. Your information is not falling out of the cosmos. It's impossible because everything came out of consciousness, existed in consciousness, and is returning.

I waited while he backed with his eyes over the board.

Me: With time, i.e. consciousness's wandering in time, it took shape, i.e. form. Consciousness created time as a way of understanding events. With time came space, the distance between events. Space and time are interchangeable distances, you know that from when they change axis in black holes. ... Let me show how information exists in different layers before I show how it was never lost.

He smiled and we exchanged a few words before I continued.

Me: What is it that determines distance? Well, the speed of light. What is the speed of light? The trick is to understand that light does not travel, but it is we who travel at the speed of light. The speed of light is a constant, yes, but what gives the constant its value for all observers? Well, it is consciousness. I can't easily show the next step, but I'm doing this backwards from an assumption. I found a new surface on the board under how consciousness created speed. With the simplified symbolic language, it would fit.

Me: So this is how I got a different velocity on c through a different area on C. If space, the perceived space, shrinks with big C, it has a side effect. Namely that small c becomes faster. When you start from scratch with this string theory, the speed of light is thus dependent on your consciousness.

We discussed and got into philosophical questions before I answered the difficulty of different consciousnesses living in parallel at different speeds.

Me: There is no problem, because as you see here they cannot interact fully as the speed c also changes Planck's constant or rather the other way around. They only see their velocity and only interact with their own matter on their plane of reality.

I paused briefly while silently pointing back to the previous equations. He was thinking at the same pace. We were in sync.

Me: If you let Planck's constant with this c now form new matter, it will not interact with the previous matter. How does that happen? Well, even the levels between the different speeds are quantified, giving them what we travelers call different planes of existence. Consciousness, in this string theory, has now formed parallel planes of existence. You would probably call it a parallel universe, a misnomer, but correct as one plane of existence is as large as the one we started from before.

I looked over the surface of the boards, and found what I was looking for. How the relations in vibrations are musical. I repeated what I had said about it and continued.

Me: I can't go through everything, but you'll have to take my word for it that this is true. If you see how the tonal relations arose, you can understand that they recur here too, between the different planes of existence. The planes of existence of the universe relate tonally to each other in sounds. And if you know how the first consciousness is a tone, then you can understand how we can make different planes meet in chords anywhere. Or how a tunnel between two consciousnesses can phase-in to each other in whole or in part. But that meeting we would call, hold on now, spiritual or a spiritual revelation!

We kept laughing and pulled some funny situations. It was needed to ground the perspectives. "Yeah, now we'd just have to space jump to this," he blurted out. "I can do that", I replied just as cheerfully. I pointed and continued.

Me: What if we remove the cause of time that consciousness created? Well, then we have a local wormhole!

We laughed together, again. I enjoyed it so much.

Me: How do I arrive at that? Well, the same way I did between two different Cs. And so I deliberately forgot to say that here we have exchanges between two different Cs. And shall we get really pseudo-scientific? (I smiled big.) Well, it's called telepathy! The transfer of information between two different C's.

Now we were having fun, as it should be, so I had him in my time.

Me: These were wormholes for information. But then how do you create a real wormhole that we can travel through. It works the same way, because you could say that your craft is a manifested information package. And now looking back at original C, wormholes don't arise by themselves, they are created. And for this, no energy is required, because there is no energy needed. Everything is done by technology, yes, but it is consciousness that guides and opens.

Here he became thoughtful. He used to tilt his head forward a little bit when he was thinking cleverly and he was doing that now. When he was more philosophical, he could nod slowly. And the few times he didn't, it was as if his body locked up, stopped by the thought. It didn't happen often and that was precisely why I observed it.

He: Hm, that would mean that more advanced civilizations would not consume more energy as we thought according to the Kardashev scale. The really advanced ones would be traveling mentally.

I waited before filling in.

Me: Yes, absolutely correct. And if you imagine that in the almost infinite number of planes of existence, little c reaches no speed at all. The big C that bears this would thus be everywhere at once. That's why you need to start at this end of string theory to get to the God Formula, which is what I call this, not the equation explaining the various forces. You've already seen that everything derives down to big C. And to explain that, I've done what I call a negative regression analysis, simply a negression.

We went back at his request to clarify that he understood.

Me: Here on the left we have a wormhole for information. But for a real wormhole for particles, with so-called collapsed wave function, requires no energy... Distance does not exist, potential barrier does not exist. Wormholes are built by C or rather cease with the absence of c. What did I write here? The distance didn't exist in the first place, so we took it out of the equation. No displacement took place, no energy was consumed. No negative energy was needed to bend space time either.

He: I'm with you, but since you haven't written out all the equations, I obviously have to check this. Over and over again I have to remind myself that you're a time traveler and I'm working on not seeing before I understand. It's obvious that you have the knowledge of the future with you here, when I see this.

Me: Yes, I do and it's through a link with my future. I'm sort of reading from the journals I haven't written yet and I'm doing it through a tunnel of information!

Finally, we laughed together again. I had been running at a high pace for a long time and it had become difficult to synchronize. I increasingly needed to rely on his pace and adjust mine.

Me: You see here that the base of the movement through space time is E, events. What is consciousness in the equation is C, outside of time, and that creates space time out of E. These E are strings in time between two poles, the two original C. To reach the God Formula you need to start where E starts and then you can write it as a string theory. I'll leave that open here. It's not a big area, but you don't need more. Should you need more surface area, you haven't captured it in its simplest form.

He looked, nodded, pondered, and nodded again. I saw that he was filling in.

Me: What I've done here is to create a thought pathway that you can build the thought structure within or rather to.

He: What can I say about how this came about? I can hardly say that it was a time traveler from the future who came with it.

Me: You won't say anything. I'm from there so I know. But what I didn't say was that you won't remember me as anything more than a thought, any thought. That's the law. I may leave a thought, but no formulas from the future and certainly no characters.

When I erased the blackboard, even though he wished to keep it, I was erasing not just one time but all. Only the waves remained, for someone to catch in inspiration. I did not want to see how it went. Such things hurt too much. 30 seconds left...

Me: SubWay is my way. Subconscious is your way!

I had done it! I had just done it! But was it necessary to erase everything? It was painful. I was not only erasing my thoughts, but also myself. The desire to become a historical person is in all of us. That is actually to become immortal. That is to become an event with a big E.

You can't erase your own tracks without erasing a piece of your self. I didn't let that influence my actions and afterwards it was too late. Inside me, I felt less real. It must not affect me, I told myself. I wouldn't make it to my death in time otherwise. I was good at keeping my focus on the future no matter what I was going through in the meantime. I was also good at time loops without breaking the laws of time. I knew the permissible time loops that strengthened time and that utilized time as a wave of E.

I was good at one more thing, but I hadn't gotten there yet. I could only guess that the path of my thoughts must be deeper than I could understand. And understand I did not, until after the time wars were over. There was my E that had propelled me through the most difficult path I had ever seen anyone take. It was my C's first encounter. I smiled, "think how the beginning lies in the future". I laughed uproariously out loud as I said, "This whole journey before I come into being so that the beginning comes into being. That's reverse causality right there!" It is fortunate that no one in the universe knows that the universe has not yet come into being. The beginning is soon.

I wrote it down then. Now I know it's true. I am, at the time of writing, in the middle of the beginning, the biggest time loop of all.

The formula for life

I knew I had touched on one more topic. But was it then? It must have been.

To find an episode in time is to find it in memory. When time is erased, memory is erased. Only the feeling remains and that was what I had now. We must have had more than our hour in the hall. How could it be so?

I wondered how he would perceive that we had only stayed there for 10 minutes. Could I have changed the time so that even I didn't realize there was more? And why not? The question then was why, and it should have an easy answer. We had got onto an absolutely important topic, that was it! A subject that just had to be given full explanation time. That was it! But what was the topic?

Finding the topic was necessary to find the document, my notes from the episode. I searched and searched, but found nothing. I was discouraged and at the same time both frustrated and annoyed. None of the emotions can exist because they steal the space for the unthought, the thought that comes as an aha experience. And it happened! Probably because I remembered the feeling of the unthought space, the feeling of the aha experience was enough to awaken the void of thought that gave birth to the unthought. The unthought was as obvious as the most natural logical thought could be. Obviously, within a time loop I had already arranged in the hall, there was only one more way to expand it. A time pocket.

A time pocket can only be discovered if you touch the surface of time backwards. I had felt that there was something more I had not accessed. It was as if I had brushed my hand across time down my coat. The lid of the pocket was not felt but a bulge from that unknown in the pocket was felt. I had done this before, during the time wars, when searching for a lost friend. I searched with my hand, stroking again and again until I knew where it was hidden inside my coat. It awakened my conviction and that's what I needed to stroke backwards through the surface of time. The process is played backwards in the body, in the same way that a hand strokes a surface a little reluctantly in the other direction. There it was, the lid of the time pocket. It is strange to go backwards in something that had not really happened, yet it had.

And just as I felt the pocket lid of time running out, I turned again, 180 degrees, so that I was moving forward. Suddenly I was in the lost time moving forward where I had crossed over and... there was the event, the topic we had discussed, the formula for life. I repeated several times until I had memorized and could take the event out of my pocket. At that moment, I pulled out the hand I had been holding in my document drawer. There was one sheet, one sheet is more than nothing. Just short notes, that's why it was so elusive. Good memories need complete notes. But there it was, the formula for life.

I just had to sit and breathe. I didn't dare let go with my hand or my eyes. It had happened before, the relaxation made me drop the caught. Last time, unfortunately, it did not fall to the floor, but continued out of time for good. Of course, I couldn't remember what the subject was, even though I had spent a local hour trying to capture the document by memory.

I hurried to start writing this down so that the document would have one more attachment point. It was needed because the mere subject itself was elusive. In the blink of an eye, my maneuver hadn't been noticed. The fact that a board was suddenly blank wasn't noticeable because we were apparently traveling in the right direction. Only after I was done would the pocket get an outside.

But how had it got a lid? It wasn't something I did deliberately, because that would be a time war maneuver when someone wants to steal something and hide the travel. It must have been because I was traveling in the same direction as the flow of time. Time tends to be self-healing so as not to cause ruptures and collapses.

Hm... We had reached the God Equation and then he just blurted out that now we would just have the equation for life too. An irresistible challenge for me. I took the opportunity to hand it over when there was an opening. It wasn't unfounded because he had touched on the subject several times when talking about the movement of thought...

Me: Shall we?

He: Dive into the formula for life, you mean?

Me: Sure!

He laughed at the pleasure of getting another round of absolutely groundbreaking thoughts. I jumped right in so that he wouldn't look at his watch too closely. I just loved his whole so vibrant being. He was in thought the discoverer of continents behind the curvature of the Earth.

Me: Okay, you know my approach is usually to explain and then formulate it mathematically. So we'll start again with the figurative language and transcribe it into the mathematical language.

Hold on tight! The definition of a consciousness, as we perceive a consciousness, is that it is not part of the universe!

He had that shaking laugh that also said that this was off to a promising start and that he was about to hear an explanation he had never heard before. I continued.

Me: A consciousness everywhere is the universe, the way I said the movement of thought moves. A consciousness can form a self, as we know consciousness, by having a focal point. If you have a focal point, then you are not everywhere and thus, in our view, not part of the universe. And then there is the question of how a consciousness with a self originates...

We reflected on some of the earlier arguments about the necessity of the observer and came back to what constituted a measurement, and I meant a consciousness that was not separate from the process but still in the universe. I proceeded to a deeper definition of self-consciousness.

Me: What gives consciousness a focal point is a specific circular movement of thought. It returns to its starting point and continues. That movement of thought gives a geometric shape as I showed earlier. The circular movement gives consciousness an isolation from the universe.
He smiled, laughing, and he realized where it was going.

Me: With each turn it has made a small change which makes it a spiral or rather a vortex. Through this we have one of the foundations of life. Life brings change and is change. But let us return to what the vortex means.

The circular movement of thought produces an inner consciousness separate from the outer. As there is no size, the inner consciousness is also a universe.
We paused in shared silence. It was so good that he could allow that.

Me: To fully understand the movement of self-consciousness, see it as a vortex, a shape, in the water. The vortex is part of the water and of the water in the water, because it is the movement of the water. At the same time, the vortex and the surrounding water are separate. Separation is a necessity for the observing consciousness.

We returned for a while to the idea that what I call the Sea of the Source, the origin of allness from which everything has taken shape, could be translated as Quantum Field, but with that came problems. By using the same term, what I describe would be confined within known limits. This term was also closer to the analogy from which I showed the origin of how self-consciousness is born out of movement.

Me: How then does the first vortex arise from the infinite sea? Imagine different random variations and spontaneous movements of two oceans, they would have to keep different vibration and thus different speed. And that is what they do. Call one warm and the other cold. What's funny is that they actually feel like they have a difference in temperature. Yes, you have to take it from a traveler. This makes you have one slow and one fast movement along an infinite uninterrupted surface, which they are. I'll give you a spoiler as a traveler, all of reality is just a surface and it is when time is added that that surface is given events and thus the depth of space. I apologize for that. I'll return.

I was getting anxious and that was not good. Hold the line, hold the line, I muttered to myself.

Me: When the two oceans move at different speeds, life as we know it arises as a consciousness that is self-aware. A separate movement from the surrounding ocean or, you know, reminiscent of Quantum Fields.

Here I began to write down the simple formulas of angular velocity as relative movements in the ocean, the relative field, but absolute to itself. I went on to explain, for a life to become eternal and self-sustaining, three vortices were required in succession or rather in parallel. I showed how each vortex caused layering by moving with ever finer motion. I pointed to the previous round on the top right board, how we arrived at Planck's Constant was one of several. He nodded, showing that he had already understood the last summation.

Me: As I showed earlier how Planck's Constant is one of several values, I have shown here that the constant thus originates from the fundamental movements of consciousness which form life. Little Omega (ω) is actually the angular velocity of consciousness (big C). And that's the one that sets the value through the formula for the speed of light (small c)!

He lit up and was more excited than ever. We went on to talk about how this would explain how light does not have a speed, but is actually a result of consciousness. The speed of light is actually the speed at which we travel through space in all directions.

He: What you have shown here, as you as a traveler mentioned, is that there are multiple planes of existence, which I now understand are the same as planes of consciousness! Ingenious!

He looked over the whole surface again as if to check that he hadn't said too much. The closer he got to the end in the bottom right corner, I saw what I had seen in those who had had an epiphany. How could I put it into words. Revelation is the best word, and with his cordiality, which I recognized, touched to the soul, it made me try to incorporate one more thing. Perfect for this moment, because I saw the light in him.

Me: Can I add? (He nodded.) With some caution we can name the three vortices. The outer one is what I call the resonant body, the first necessary body which everyone needs to contain a separate consciousness. You can call it the body of matter. And the second one is frequency-wise like a content to the first so-called container, you can call it the inner secluded water of the sea of beingness, the Quantum Field – the Soul. And what about the third? Well, it is how air relates to the "density" of the previous two. You'll recall that I called this mathematical weighing above density. The word for this swirling distinction becomes Spirit.

He looked further at the board. It was a little more difficult this time, but he had a stubbornly open mind. So I was able to get to what would make his thoughtful face break out smiling again.

Me: Let me take time out of the equation here, I divide on both sides. And what has this done? Well, everything is flat again. Timeless, everything is surface. But... Ready? (He nodded, knowing.) If you instead put this in a two-way causal event perspective, you suddenly have a string of something we call slow big Omega and fast big Alpha. The events are connected. Through time and space. But without time they are one and the same. What consciousness does is it adds different speeds and gives us continuity, after the body has fallen away. Yes, I'll say it. It is not possible to give you the formula for life without showing that it is the only thing lasting through time. The Soul and the Spirit are eternal. Here you have the end of string theory in the way you would have to start.

It literally bubbled out of him, literally everything between heaven and Earth. That the path to the God formula really meant God formula with this ending. And the formula for life was really the formula for consciousness as observer and thus creator of events. "In the image of God, I added". It was all obvious how this ultimate omnipresent Quantum Field was consciousness and together (Alpha and Omega) were the origin and sustainer of everything.

I nodded and unnoticeably I packed a little. A piece of paper first into a book, then a book into the bag without looking at it. He followed the train of thought I laid out. With relief, I thought that he would never notice that I had made a time pocket. With a certain emptiness, I thought this would hardly stick. Maybe, just maybe, if time, with the help of his thoughts, got closer to the outside of the pocket, a "new thought" would be awakened that would give him a revolutionary path, a restart in his string theory building. It was painful, time- traveler-painful. Why do I put myself through this, time after time? I smiled, he didn't see it. He chattered on all the way until it was time to part ways. I pretended to have to take the stairs to exit.

I took a hesitant step in time, I wasn't quite ready. I stopped and sat on the park bench outside twelve minutes before I was supposed to go in to his class. I had learned to tie time so it made it easier to accomplish what I wanted and needed within the limited time opening. When it was time for me to be in the classroom, I had already left. Overlapping myself only caused problems. Just before, I remembered that this would be difficult to leave in time for myself, but it was too late, it was tearing and I was already on my way.

Memory is as elusive as time.
Time is as elusive as memory.

The written word therefore
writes time.

Without a body, there would be no
self.
A life as mere consciousness does
not exist.

One sea has no movement.
Two has.

Without The Two seas of consciousness
in their movement towards each other
self-consciousness cannot arise.

The movement from low frequency
and rotation to higher finer, is called
evolution.

But few understand that in the movement
called evolution
the distancing from the beginning occurs.

For those in time Alpha,
Omega departs.
For those in the original meeting,

only seemingly.

The distancing gives evolution
and space
in thought and time.

So does the approach.

Alpha cannot move further away
than it loses the Omega in itself.

Without Omega in Alpha,
the movement collapses.

The return is inevitable.

No movement starts without its
completion.

It was completion
that made the movement start.

Can a separate self-consciousness be
greater than its sea?

Yes, when the self-consciousness
understands
that it bears within it a universe.

Understand a sea,
understand a consciousness.

Understand you,
understand the movement of your
thought
that separates.

To understand a sea,
to not understand your consciousness.

To lose the thought that
separates you,
to lose your focal point.

Understanding a Sea
as a Sea.

Embrace with your heart
Everything and Everyone.

Your inner Sea as your surrounding,
All One.

Understand a Sea.
Understand another.

Fall into eternity.
Fall in love.

This is how The Two bearers bear
themselves and each other.

Out of eternity.
Into eternity.

How does a Sea understand
another?

They don't.
They fall in eternity before each other.
They fall in love.

Fall like that,
like The Two.

The End of Future

He: When will I see you again?

Me: Something has come up. You ask because you know what I know.

He: Explain.

Me: You have a gut feeling that I won't be back. That's how events in time reveal themselves about what's to come.

He: You mean I can sense such things?

Me (smiling): Of course you do. Every day your life is changed by what is to come. Every day you make decisions based on what is to come. Everyone does that. Do you remember my first words to you? Reverse cause and effect exists. We had to meet because of the effects. This moment now paved the way to our first moment.

He: I am shocked. Again. I don't understand it, and yet I think I do.

Me: I don't use a machine to travel. You haven't asked because you know. You know in a way, you don't know how, but know. This is the mind timeless.

His breath told me that he was stunned again.

Me: I answer you without you asking. Yes, you fall into the empty time-mind-space. Yes, we are billions of years old and our minds utilize time. We all have our limits. I have reached mine. My timeline is broken.

Chills went through the room. His chills. He knew, but not how he knew. Too late to describe how emotions create the bridges of time.

Me: I promise I'll try to return before we met, to save our moments.

He: Why? I don't understand.

Me: Death breaks a timeline. But to save lives, I must change our moments. You already know. They will not exist anymore.

He looked at me and asked "Will we never have met?"

Me: I'm sorry. In time we will not. But outside of time, memories will endure. If I can, before I die, I will try. I will loop myself, and if I manage to remember myself, I will be able to return to you with the memories of us. Please memorize this. please do. If you do, you might remember.

I wrote him a note:
Quotes from the Journals of a Time Traveler.

His mind had grasped the unfathomable.
I left, he watched me go.

The very beginning

Me: Excuse me! Sorry to interrupt you but can I have a quick word with you?

He: Uh... I'm in the middle of something right now and... Can I talk to you later or can you wait until after my lecture?

Me: Please, do you have a minute of your time? This is my very last time here and I can't come back.

He: I'm sorry, have we met before?

Me: Yes several times, in the future. It is an honor to meet you. I am a time traveler.

He looked at me with curious eyes and a little reserved.

Me: It will be short. I have to return very soon, my window of opportunity is closing as we speak. You can't remember me yet. But maybe if I give you this book. Here are quotes from our moments we had in the future.

I couldn't hold back my tears. I continued in a slightly forced voice.

Me: I am so sorry but I was killed. The only way I could save our moments was to write them down and hopefully rediscover them again to get them published. As I said the first time we met: Reverse cause and effect exists. We had to meet because of the effects. Please take this book. Put it in your pocket. Do not drop it for the first few hours. It may not be able to stay in this time. If it is gone, so am I, and so are our moments in the future that now become parallel.

He found it hard to accept my words but the tears on my cheeks spoke louder.

Me: It has been an honor to meet you.

Just as I turned away, he just had to ask me.

He: Is this a joke?

Me: No. I wish it was.

I left unnoticed. In the hallway no one saw me go out. Just as I had told him about going in:

Where I go into time, I go into consciousness. Therefore, I enter your mind, and not with lightning. You will think of something else and will not notice me appearing.

I never knew if he kept his promise not to let go of the book, or if I was erased. I didn't know if he would remember the note I had written to him, Quotes from the Journals of a Time Traveler.

I had one more mission, to hide my journals, the impossible mission. Time was the only hiding place for my journals, to keep my drawer empty. If I survived, would I remember how to retrieve? Remember time that had not happened yet and no longer existed?

The final post to him:

Now you realize that I wrote this for you to remember the abilities you once had.

Become who you became!
One more time!

I know you can do it!

Published books and booklets by Mikael Kvist

Updated 2026

Mirjams Bok (1992)
Möte med Mirjam (1992)
Marias Röst (1994)
Ärkeängel MiChaEl (1995)
Drottning av Kristall (1996)
Tidkrigarna I, II (1998)
Jordens Kristallnät (1998)
Kanalisering (1998)
Kort och Gott om Knytt (2000)
Naturväsen (2001)
Healing (2003)
Dimensionell Årsbok 2020
Så som änglar vill bli förstådda (2021)
En Utomjordisk manual del 1 (2021)
Utomjordisk manual del 2 (2021)
Dimensionell Årsbok 2021
Mellan möten och uppdrag (2022)
Dimensionell Årsbok 2022
Portaler till parallella världar (2024)
A Manual for visitors of Earth Part 1, 2 (2025)
En Utomjordisk manual del 1, second edition (2025)
Utomjordisk manual del 2, second edition (2025)
Citat från en Tidresenärs Journaler (2025)
Quotes from the Journal of a Time Traveler (2025)
Tidkrigarna I, II, second edition (2025)
Time Warriors I, II (2025)
Kort och Gott om Knytt 1, second edition (2026)
Dimensionell Årsbok 2023 (2026)
Dimensionell Årsbok 2024 (2026)
Kristallportar I-III (2026)

Plattform for The United Fleet on Patreon created by Mikael Kvist 2022: https://www.patreon.com/unitedfleet

Web page for The United Fleet: https://theunitedfleet.one/
Web page for Mundum: https://mundum.org/

About the author

Mikael Kvist is the author of over 20 books and pamphlets in Swedish. Mikael had a technical background as an electronics engineer and studied paranormal phenomena since the mid-1970s. Mikael had over 35 years of experience studying the Earth's crystal grid and its portals. His clairvoyant studies led to unconventional methods of technically measuring and studying paranormal phenomena. The unique research results were presented on the PARCS association's website, i.e. the PAranormal Research Center of Sweden's website at the time.

During the 1980s, Mikael was editor of Vita Vingar, among other publications. Until 1999, he wrote several series of articles for the magazine Alphaomega. These caused quite a stir due to his insightful and critical view of channeling. He himself was a contactee for The United Fleet, which meant that he was in constant contact with beings from other worlds.

Mikael held classes for over 15 years. Among other things, he taught Multidimensional Psychology and Angelic Love. He also lectured on topics such as the Earth's crystal grid, paranormal research, channeling, nature beings, etc. He was in high demand for his clairvoyant experiences and coaching.

Mikael Kvist (1960-2024)